eBook Bonus.

With the purchase of this book, you also receive access to its electronic version. Use the following personal code to download the electronic pages.

9r62l-guph4-
01800-at1bx

Take advantage of the eBook Bonus on your computer, smartphone, or tablet. Go to **www.hanserpublications.com/ebookbonus** and register with your personal access code.

Download of this book as an eBook is subject to legal regulations, which can be found at www.hanserpublications.com/ebookbonus. Requirements: Internet connection, Adobe® Reader®, and account registration.

Bhargava

Robust Plastic Product Design:
A Holistic Approach

Vikram Bhargava

Robust Plastic Product Design: A Holistic Approach

HANSER

Hanser Publishers, Munich

Hanser Publications, Cincinnati

The Author:

Vikram Bhargava, Somerset, NJ, U.S.A

Distributed in the Americas by:
Hanser Publications
6915 Valley Avenue, Cincinnati, Ohio 45244-3029, USA
Fax: (513) 527-8801
Phone: (513) 527-8977
www.hanserpublications.com

Distributed in all other countries by:
Carl Hanser Verlag
Postfach 86 04 20, 81631 München, Germany
Fax: +49 (89) 98 48 09
www.hanser-fachbuch.de

Cataloging-in-Publication Data is on file with the Library of Congress

Editor: Dr. Julia Diaz Luque
Production Management: Jörg Strohbach
Coverconcept: Marc Müller-Bremer, www.rebranding.de, München
Coverdesign: Stephan Rönigk
Typesetting: Kösel Media GmbH, Krugzell
Printed and bound by Hubert & Co GmbH und Co KG, Göttingen
Printed in Germany

ISBN: 978-1-56990-580-7
E-Book ISBN: 978-1-56990-581-4

Dedicated to

Mother Saraswati,

my parents, who dedicated their lives to the cause of education,

my wife Pushpa, who lovingly allowed me to steal a few precious retirement years from her in the writing of this book,

all those who encouraged and cheered me on to write,

and to my alma mater, Birla Institute of Technology and Science, Pilani, India, for shaping my personality to be a productive engineer and instilling in me the fundamentals of engineering that were my guideposts from my first day at work to the day I retired.

Acknowledgments

Tim Austin, who moved from England to join me in setting up the plastic flow simulation resources at Symbol, and for suggesting the term "holistic approach" for my training seminars.

Glen Beall, whose training and training videos going back to the early eighties were the inspiration for my own educational seminars–for going through the manuscript and for writing his foreword.

John Beaumont, for his guidance in the section on multicavity molds.

John Bozzelli, for his limitless passion in promoting scientific molding and introducing me to the technique.

Branson, for allowing the use of various figures and data from its resources. In particular, Jeffrey Frantz, who has helped me deal with extraordinary welding challenges over the years.

Covestro (formerly Bayer), for allowing the use of various figures and data from its resources.

Julia Diaz Luque, for her patience and skills in the editing of the entire book and "keeping me honest."

Dukane, for allowing the use of various figures and data from its resources. Also for sharing knowledge on round energy directors.

Suhas Kulkarni. The plastic industry owes him much for his one-man crusade in making Scientific Processing a "household word" in the processing community. He has contributed richly to the processing chapter.

Louis Maresca, for going through the manuscript and writing a foreword.

Joe McFadden. An engineers' engineer. There was no failure too big or too complex for him to tackle. I have learned much from him. Also for helping with various sections in the book, including a foreword.

Motorola, Inc., Motorola Solutions, Inc., and Symbol Technologies, Inc. I have learned much from my participation in their relentless quest for the creation of ultra-robust, customer-focused products.

Mukund Patankar. THE authority on polymer basics, for helping me with various sections in the book. His participation in my group brought so much knowledge of polymers and other areas to Motorola.

SABIC, for allowing the use of various figures and data from its resources.

Prachi Sabnis, for creating numerous diagrams from hand sketches.

Jagroop Singh. Our man in Hong Kong and China, my bilingual co-trainer in Asia and the US—for all his help in the tooling section.

Mark Smith, for his professionalism, coaching, and playing the role of a cheerleader in the development of this book.

Sonitek, for allowing the use of various figures and data from its resources in the book.

Weng Onn Tang, for his inexhaustible energy in keeping on top of projects in China and the translation of documents from Chinese to English.

Mark Yeager (Covestro). I have learned so much from him in the proper use of materials, design, tooling, and processing. His work at Bayer and Covestro in creating design books, white papers, and seminars has been a real boon to the plastic design community.

Finally, the countless professionals, business associates, and training attendees, for their encouragement and support in the conception of this book.

Preface

Over the years, my colleagues and coworkers have both appreciated and derided me for my "Vikisms" (a term they coined). One of the ones that I am not embarrassed about is:

"Knowledge is knowing that one does not know."

Unfortunately, when I graduated with a Mechanical Engineering degree (albeit from a prestigious institution) I did not have this "knowledge" in the field of plastics. I thought that I could create successful plastic parts by using the assumptions and structural calculations that worked almost without fail for metal components. In addition, I knew little about the complexity of polymeric materials and the effect of tooling and processing on the cost and performance of parts made from them.

I remember the first miniature switch housing I designed for a major U.S. leading technology corporation.

The material chosen was polycarbonate. Poor choice for this application, as will be shown later. The main wall was about 0.5 mm. The internal ribs were about 2 mm. When the part was molded, there were huge sink marks under the ribs and the part had extreme distortion because of the resultant uneven rate of cooling.

I was also responsible for the tooling. I located the gate at the thin wall, for I did not know any better. The part became very hard to fill.

As the molding supervisor, I raised the temperature of the mold and the melt to fill it more fully. Strange—that made the sinks and distortion even worse! Over the course of several days, I tried tens of combinations of mold and melt temperatures and pressures, with little success. Finally, through trial and error I was able to keep the part somewhat flat and with less obvious sinks by running the mold very cold and keeping the material very hot.

The switch was designed to be used in an automotive shop, among other environments. It developed major cracks the first time it was handled by an operator with grease on his hands, due to the internal stresses caused by the poor choice of molding parameters. (Even if the part had been well designed and processed, it was doomed for ultimate failure in this environment because of the poor chemical resistance of the material to grease.)

Seem implausible?

Just look at some of the examples in the book, including the failure of an automotive windshield wiper lever. The failure of this lever during a downpour could easily have caused death and destruction! The part was probably designed by an engineer who had been designing successful machined or forged metal parts all his/her life.

Over my working life, I have learned much about plastics. Some of this from making mistakes. Other aspects from attending conferences, seminars, reading books, working with resin suppliers, toolmakers, and molders. But mostly from creating and conducting training seminars. There is no better way to learn than to teach. In my initial years (and even now, occasionally) I was (am) asked questions that I did (do) not have answers for. Going back and finding answers for these questions has been the best teacher for me.

What is the ultimate role of an engineer? To help the business become more profitable by creating products that the customers love and that are manufactured in the most efficient and cost-effective way. I have been fortunate in having worked for Motorola, the company that came up with the Six Sigma methodology. I also have had the opportunity to use this methodology for the overall improvement of not only the products but also all those who were involved in the development and manufacturing of the same. It is my hope that the brief chapter on Six Sigma will generate enough interest in the readers' minds to explore it more.

Finally, to paraphrase an old cliché, this book is written by an engineer, for an engineer. I have worked in the trenches long enough to realize what a pressure cooker environment an engineer works in, with the breakneck pace and competitive nature of businesses today.

There are many books on plastics written by scholars much more knowledgeable than me. However, a good many of them are either highly compartmentalized into polymers, design, tooling, molding, etc., or are too complex for everyday need. This book is intended to provide just enough information for the engineer working under the unrealistic deadlines of today to get the product right the first time, every time.

No more, no less.

The other unique feature of this book is its holistic approach. No more compartmentalization.

I owe much to many for helping me live out my assumed productive engineering career. The dedication and acknowledgment sections detail all the helping hands along the way.

Vikram Bhargava

Somerset, NJ, U.S.A.

August 2017

About the Author

Vikram Bhargava is a Fellow of the Society of Plastics Engineers and Past Chairman of its Product Design and Development Division.

In 2014, he retired as the Director of Mechanical Engineering Services at Motorola Solutions in Holtsville, NY heading an international group of professionals overseeing plastic material selection, design, analysis and simulation, tooling, and processing.

He has over 40 years of experience in product design, development, manufacturing, and management, especially in plastics.

Vikram is a sought-after international trainer and has trained thousands of engineers and suppliers in the proper design and manufacturing of plastic parts and assemblies. Some of the companies where he has conducted extensive training seminars are Motorola, nVidia, Hunter Industries, Google, GE Healthcare (U.S.A.), Foxconn, Wistron (China, Taiwan), Geometric, HCL, and GE Healthcare (India).

He has presented dozens of papers at the ANTEC (SPE) and other technical conferences. One of them was judged as the "Best of ANTEC 2011."

He is an active blogger and authored numerous white papers and articles.

In addition to his management responsibilities, Vikram chose to remain a hands-on mechanical engineer until his retirement and designed or supervised the creation of hundreds of innovative products.

He holds 21 U.S. and international patents.

Vikram has successfully led and mentored many multimillion-dollar Six Sigma projects and is a certified Motorola Six Sigma Black Belt and trainer.

He has a BSME from Birla Institute of Technology and Science, Pilani, India.

He currently lives in Somerset, NJ, U.S.A. with his wife Pushpa and is actively consulting and conducting training and seminars in plastics and Six Sigma.

LinkedIn: *www.linkedin.com/in/bhargavavik/*

Foreword by Glenn Beall

The industrialized countries of the world depend on manufacturing to provide jobs, grow their economies, and generate tax revenue. Manufacturers can only produce what engineers design. Engineers need to know how to design new products. If there is no design, there is nothing to manufacture, jobs are not created, and there is no profit to be taxed. New product design is important to the world's economy.

Plastic is the third largest manufacturing industry in the U.S.A. In spite of that, the majority of those receiving technical degrees are taught little or no plastics technology. Even fewer receive instruction on plastic product design. Despite that failing, there is a steady stream or flood of new plastic products entering the stream of commerce. How did those engineers learn to design those plastic products?

They learn by attending technical conferences and expositions. Others learn by joining technical societies such as the Society of Plastics Engineers (SPE). Some learn in the school of hard knocks by the expensive and time-consuming process of trial and error. They also learn from the internet, technical journals, and books. And that brings me to the purpose of these comments.

I started designing and developing plastic products in 1958. The word "plastic" was not spoken at my University. I learned from all of the above sources. As a college graduate I was accustomed to learning from published literature and books. Fortunately, I had some excellent mentors who knew the difference between textbook theory and real life.

Over the years I have read every English-language book I could find with the words design and plastic in the title. I just recently read the galley proofs of Vikram Bhargava's "Robust Plastic Product Design" book. I was pleasantly surprised to discover that it is not just another rewriting of the already published "Plastic Product Design Technology." All book authors, myself included, absorb what is already known and add what we have learned in the course of our careers. If those careers are long and rich with actual experiences, the resulting book will benefit those with less experience. "Robust Plastic Design" is that type of book.

Vik has had a long and rich career. He joined the plastics industry in 1968. His first job after receiving his degree included being an injection molding machine operator at an original equipment manufacturers captive molding shop. It is a rare graduate design engineer who possesses that valuable hands-on experience.

Following 43 years spent working in large corporations, he formed his own plastic product design and failure analysis consulting business, which is still active today. He has now been designing and developing plastic products and parts for 46 years. Much of what he learned from all of those projects is included in this book.

I met Vik in 1990 when he joined a small select group of designers who were forming SPE's Product Design and Development Division. He served on the Technical Program and Newsletter committees before being elected Chairman of the Division for the 2005–2006 term.

Many years of contact with design engineers sensitized Vik to the general lack of plastic product design knowledge in the industry. That realization resulted in his creating and teaching product design and failure analysis seminars, starting in 1995. The old adage that the best way to learn is to teach is very true. Vik's teaching has added to his understanding of the type of problems faced by the product design community. That understanding also resulted in his book, which is another way of teaching.

Vik's book covers all of the usual product design subjects without getting bogged down with the scientific mumbo jumbo surrounding polymer science and molecular structure or the nuances of the increasingly complex scientific injection molding processes. He then goes on to discuss other less frequently included subjects, such as design related consideration for secondary operations, assembly, part and tooling cost, quality control analysis, improvements and part failure, cause analysis, and a brief but good review of coloring plastic materials. All of these subjects are backed up with the most comprehensive list of reference and websites that I have ever seen. This extended list of plastic product design related subjects accounts for the book's "a holistic approach" subtitle.

"Robust Plastic Product Design" is a little different from the design books I am familiar with. It covers more than the basic plastic product design subjects, which is good. In my opinion, this book will be useful to anyone who designs plastic products and parts or who interfaces with plastic product design engineers.

Glenn L. Beall

Libertyville, Illinois, 2017

Glenn Beall, an engineer, consultant, educator, and editor, has been addressed as "a one-man education crusade" by Plastics News. He has taught over 30,000 people at more than 770 seminars and has 35 patents. Beall received the award for outstanding achievement in plastics education from the Society of Plastics Engineers (SPE), which cited his work as an instructor on design and other aspects of technology. In 1997, he was inducted into the Plastics Hall of Fame in recognition for his outstanding contributions to the plastics industry.

Foreword by Louis Maresca

Vikram Bhargava's new book, entitled "Robust Plastic Product Design: A Holistic Approach," is not only well written, informative, and practical but comes at a critical time, as the United States gears up to regain its position as the preeminent manufacturer in the world. The United States has long thrived on its ability to manufacture things and sell them in global markets. However, in recent years China has surpassed the United Sates as the leading manufacturing country in the world, a title the United Sates had held for over 100 years. This has led to serious concerns about our economic future.

1. Manufacturing is essential to providing a strong foundation for economic growth. It provides the opportunity for high quality and good paying jobs for American workers. The impact of a healthy manufacturing sector also has a ripple effect on the economy. On average, each manufacturing job supports 2.5 jobs in other sectors and, on the upper end, each high-tech manufacturing job supports 16 others.
2. Manufacturing also serves as an engine for innovation and knowledge production. Historically, the manufacturing sector has been tightly linked with the nation's R&D activities, and a strong manufacturing sector that adapts to and develops new technologies is vital to ensure U.S. leadership in innovation.
3. Domestic manufacturing capabilities using advanced technologies and techniques are also vital to national defense and homeland security.

One of the areas most affected by the U.S.'s outsourcing of manufactured goods is the plastics industry. In the past, the U.S. was the leader in the technology and art for the design and production of plastic parts. However, as the manufacturing base moved offshore, so did the know-how. Today, most plastic parts are built in Asia, where much of the technology now resides. Since plastic part design and manufacturing is as much an art as a technology, rebuilding its base in the U.S. is not a simple task. Production of quality plastic parts is a complex process that requires a fundamental understanding of material properties and processing, part design, and customer needs from both an application and a costs perspective. Bhargava's new book addresses these components to provide a holistic approach to design high quality parts that can be manufactured quickly the first time.

In my 35 year career in the plastic industry, I have witnessed many examples where overlooking even the smallest component of the plastic part manufacturing process has led to the failure to commercialize and even the loss of an important customer. For example, a very large customer requested the development of a material that met certain properties parameters, including a strict requirement on impact strength. A product was quickly designed, which in the laboratory was a perfect match for the customer's needs. However, the first major customer production run of actual parts was a total failure. While impact strength was an important component for the longevity of the part in its application, compressive strength, which was not an initial design parameter, was critical to the success of the final assembly process. Fortunately, the product could be quickly modified to meet both the impact and the compressive strength requirements for the material and was commercialized in a timely fashion. However, if we had taken a more holistic view of the product requirements, a major crisis could have been avoided.

A holistic design approach greatly improves not only the probability of commercialization but also that of manufacturing it right the first time. I do not know how many times a tooling design flaw has led to the inability to produce completely filled parts meeting property and dimensional requirements and stability while being totally void of blemishes or stress cracks. In these instances, a holistic approach for better designing the tool could have avoided the rejected parts and the costly time and effort of revising the tool design. Getting it right the first time is especially critical today, as shorter product life cycles and customization have also led to the need to quickly modify existing tools or build new tools that meet the demands of a rapidly changing marketplace. Speed is also essential in the maintenance and repair of tools to limit unproductive downtime and loss of production capacity. If tools are properly designed using the holistic approach described in Bhargava's book, companies can create a competitive advantage for themselves and flourish in the marketplace.

I highly recommend Bhargava's book for anyone whose company wants to increase its new product commercialization success rate but especially companies in the U.S. who are struggling to regenerate the knowledge base that made it a world leader in plastic parts manufacturing.

Dr. Louis Maresca

Somerset, NJ 08873, September 2017

Dr. Maresca has over 35 years of experience in the plastic industry and has held executive level positions with global P&L, operations, and technology leadership responsibility. Dr. Maresca has a PhD in Organic Chemistry from Columbia University and an MBA from the Weatherhead School of Management at Case Western Reserve University. He is an inventor on 62 U.S. patents in plastics. Among the various executive positions he has held are General Manager of Technology at GE Plastics, Vice President of Technology at BF Goodrich, President of the Specialty Chemicals Division of Great Lakes Chemical Corporation, Vice President of Operations at GLS Corporation, and President and COO at Argosoy International, Inc.

Foreword by Joe McFadden

More than 10 years ago, working as an independent consultant, not looking for life in the corporate world, I was presented with the opportunity of being part of a small engineering team within Symbol Technologies. The opportunity came to me through a good friend and longtime colleague, George Nelson, of Stonel Associates. After meeting the Director of this team, Vik Bhargava, and seeing his passion for the sciences and endless energy coupled with his thirst for knowledge, I knew I had found a place where I could continue to grow intellectually and utilize my skill sets in an effort to prevent or analyze the root causes of product failure.

One of the many things that impressed me regarding Vik was that he was a true working manager. His deep knowledge and hands-on experience within the plastics industry was not something he would brag about. Instead, his passion was to share all his hard-fought knowledge. Vik, I found, was a natural educator.

Vik's team was comprised of subject matter experts covering key aspects of the plastics industry. He developed and promoted his concept of holistic design, solving complex engineering problems by being able to step back, look at the whole picture, then focus in on a solution.

As an educator, Vik conducted many internal and external seminars sharing his knowledge, which impressed and aided both young and seasoned engineers.

A tell-tale sign of Vik's efficacy while at Motorola is when I today hear the language of holistic design being spoken and put to practice every day. His legacy is that both the engineer and the product benefited from his tenure.

It is my great pleasure and honor to write this foreword, knowing that this book provides what I see as often missing within the engineering world–appreciation of the big picture. I am sure it will help the reader, no matter what level he or she is at within the industry, to have a better understanding and ability to both prevent and understand root cause issues that are often faced.

Joseph P. McFadden Sr.

Stratford, CT, September 2017

Joe currently heads the Mechanical Engineering Analysis and Services organization at Zebra Technologies (previously Motorola) in Holtsville, NY. He oversees the work of hundreds of mechanical engineers internationally and makes sure their designs meet the rigorous standards of Zebra, known for its robust products. He has a Master's Degree in Mechanical Engineering from the University of Bridgeport and continues to teach graduate students, courses on strength of materials and fracture mechanics, on a part time basis at Fairfield University. As an expert witness, he has won life-and-death litigations involving the failure of plastic products.

Contents

1 Introduction

1.1 Causes of Plastics Failure

In his book *Failure of Plastics and Rubber* [1], David Wright conducted an analysis on 5,000 failed parts and came up with the causes for the failures shown in Figure 1.1.

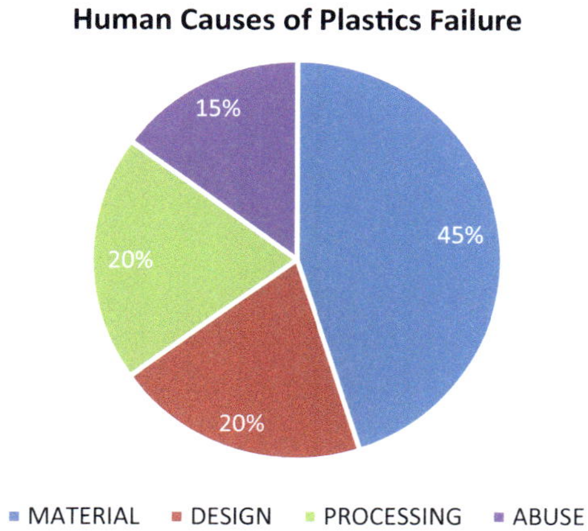

Figure 1.1 Causes of plastics failure [1]

Therefore, a designer may conclude that he or she is responsible for only twenty percent of the failures. As will be shown in the following chapters, what appears as materials, processing, tooling, or misuse issues may be traced back to fundamental design issues that may manifest themselves as the former.

Let us assume for the moment that design issues do cause only twenty percent of the failures, and the designer proactively makes sure that this 20% has been reduced to zero. Even then, in order to be certain that the product meets all the functional and cost requirements, **someone** needs to take responsibility for ensuring that the other three areas are optimized. The presence of designated materials, tooling, and processing specialists in the organization notwithstanding, that someone is still the designer!

1.2 The Holistic Approach

Let us look at the classical definition of the word "holistic" [2]:

holistic

adjective ho·lis·tic \hō-ˈlis-tik\

1: of or relating to *holism*

2: relating to or concerned with wholes or with complete systems rather than with the *analysis* of, treatment of, or dissection into **parts**; *holistic medicine attempts to treat both the mind and the body*; *holistic ecology views humans and the environment as a single system*

In the case of a well-designed plastic part or assembly, the **parts** are:

- Material
- Design
- Tooling
- Processing
- Secondary operations

The analogy I have used over my years of conducting training on this subject is that material, design, tooling, and processing are like the four wheels of a high performance car.

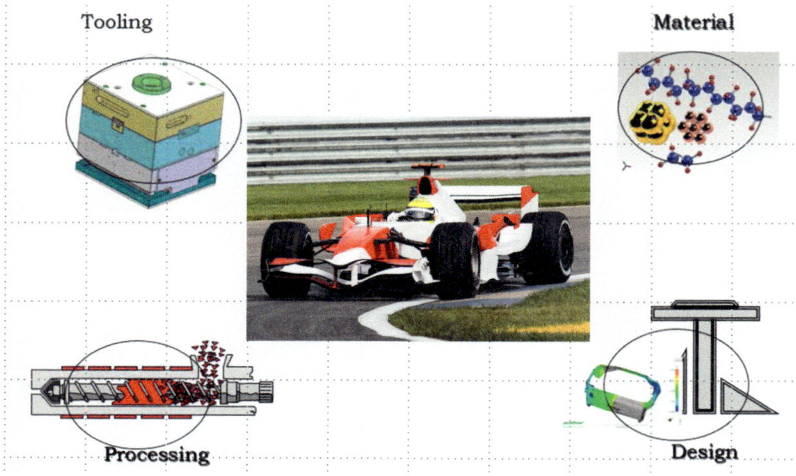

One bad wheel and ...

On the other hand, if all of the four wheels are taken care of, one can have a product that can last for years and withstand a lot of so-called "abuse." During my training sessions I often show the *video*[1] of a handheld terminal[2] with proper consideration to the four "wheels." In the video this terminal is tossed to the high ceiling of a warehouse multiple times, dropping back each time on a hard concrete floor. It is also kicked around and used as a hammer on the concrete floor. At the end of the video, there is no crack or break and the terminal functions normally.

---

[1] *http://www.vikpedia.org*

[2] MC9200–a rugged mobile computer from Zebra Technologies, formerly Motorola and Symbol Technologies. This terminal was first introduced by Symbol Technologies about the year 2000. After having been in service for over seventeen years, it can still be seen totally operational in stores like Home Depot, Lowe's, etc.

Figure 1.2 Symbol/Motorola/Zebra terminal MC9200

1.2.1 The Four Wheels

The following list contains most, but not all, of the considerations a designer needs to take into account in order to create a robust product.

1.2.1.1 Material

- Basic physical properties
  - Static: mechanical–tensile strength, toughness, elongation, etc.
  - Dynamic: wear, modulus at high impact rate, fatigue
  - Electrical: dielectric strength, resistance, etc.
  - Long term: color retention, embrittlement
- Environmental properties
  - Chemical resistance
  - Color and appearance
- Thermal properties
  - High and low temperatures
  - Temperature cycling
- Agency requirements
  - UL
  - FDA

  - RoHS
  - REACH
  - Country specific requirements
- Molding properties
  - Flow: MFI, flow length vs. thickness
  - Bonding between materials (overmolding)
  - Directional shrinkage

1.2.1.2 Design

- Purely physical requirements
  - Mechanical–tensile strength, toughness, elongation, thermal properties, etc.
  - Electrical–dielectric strength, resistance, etc.
  - ESD (electrostatic discharge)
- Molding and tooling dependent requirements
  - Gate size, appearance, and location
  - Knit line location and appearance
  - Wall thickness uniformity and wall thickness ratio
  - Thick to thin flow
  - Sharp internal corners
  - Knit line
  - Orientation
  - Draft
  - Boss lengths–unsupported cores

1.2.1.3 Processing

- Material drying
- Residence time
- Material heating
- Shear rate
- Injection time
- Packing time and pressure
- Gate freeze
- Mold heating and cooling
- Cavity balancing
- Venting
- Cooling

1.2.1.4 Tooling

- Runner and gate
- Flow balancing
- Cooling
- Ejection
- Multicavity balancing
- Venting

1.2.2 Assembly and Secondary Operations

Few plastic parts are complete as molded. In most instances, they require additional operations such as:

- Ultrasonic assembly
- Heat staking
- Insert molding
- Screws
- Silk screening
- Pad printing
- Hot stamping
- Painting

In the following chapters we will look at the details of all of the above.

1.3 References

[1] D. Wright; *Failure of Plastics and Rubber Products*, Smithers Rapra (2001) p. 6

[2] Merriam-Webster Online Dictionary; *http://www.merriam-webster.com*

2 Plastic Materials

2.1 Introduction

As discussed in Chapter 1, according to David Wright, material failures account for 45 % of the failures in plastic products. It is, therefore, important to get a good understanding of this "wheel" as a start. One of the themes that I am going to repeat again and again is that *plastics are not metals*. For most of us, our formal engineering education has been mostly in the area of metals with very little emphasis on the distinction between metals and plastic materials. We, therefore, tend to count on the same properties that we expect from metals when designing plastic parts.

More on this distinction later. Let us start with the structure of plastics.

2.2 Polymer Fundamentals

The chemical name for plastics is polymers [2]. Polymer (poly-mer) simply means many mers. Mer is derived from the Greek word meros, meaning a part or unit. A mer is the repeating unit of any polymer. One mer is a monomer, two mers form a dimer, three a trimer, and so forth. A great many mers form a polymer.

The commercial resins in use today are composites of substances with polymeric material being the main component. Additives such as colorants, flame retardants, internal lubricants, processing aides, fillers, or other reinforcement materials, just to name a few, can be found within the resins. Although the additives play an important role in the processing and final performance, we will focus here on the main component of plastics–the polymers.

2.2.1 Polymer

As stated before, a polymer is a long chain molecule which is often made up of repeating chemically distinct units called monomers. The length of this chain and its molecular weight depend upon the number of linked, repeating units and their corresponding molecular weights. In getting linked the molecules are chemically bonded to each other in that the bonding involves the manipulation and sharing of valence electrons, creating a covalent bond. This type of bond is listed as a "strong bond," as opposed to weak ones, which will be discussed in 2.2.3.2.

Within materials that we commonly use there could be tens of thousands of individual monomers in a single chain. Even with this number of molecules, the physical size of a typical individual polymer molecule is around 2,700 nanometers—well beyond the capabilities of optical microscopy. Therefore, in order to form a physical product many, many individual polymer chains are required.

2.2.2 Polymer Chain

It is the interaction of these multitudes of individual polymer chains that truly defines the plastic's cohesive properties as well as the response to external substances and stimuli. Considering a single long chain molecule, its ability to take on certain conformations depends upon the flexibility of the bond angle between each repeating unit, size of the pendant molecule, as well as other complexities beyond the scope of this discussion.

We know that there is a certain attraction between all matter, and if there is enough thermal energy, the individual long molecule will "wiggle" around and entangle itself. Knowing that the molecule cannot completely fold back upon itself, there will be space between the chain segments.

Now imagine heating millions of these chains, causing all to wiggle around taking on a number of conformations and entangling with each other. We then cool the mass, which reduces the movement, leaving behind a cohesive structure that contains a significant amount of open space. This open space is termed "free volume."

The cohesiveness of the material therefore depends upon both the level of entanglements as well as the intermolecular attraction forces between the individual molecules. These intermolecular forces are called the van der Waal forces or "weak" forces, as opposed to the previously mentioned covalent forces which form the individual chain. As one can imagine, the closer the molecules are to each other, the stronger the accumulated "weak" forces or cohesiveness of the material.

There are two broad categories of plastics—thermoplastics and thermosets.

2.2.3 Thermoplastics and Thermosets

At the highest level, the melting and resolidifying of the thermoplastics is a physical change and can be reversed. Thus, when a thermoplastic gets heated to the point where the chains start to slide against each other, it turns into a very viscous melt that can be pushed into areas that will give it the final shape—as in injection molding. There is no chemical reaction taking place in this process.

Thermosets, on the other hand, go through a chemical change in the process of heating and the chains get cross-linked through the sharing of electrons. This process cannot be reversed by reheating the cross-linked chains.

2.2.3.1 Thermoplastic Types—Amorphous and Semicrystalline

As explained before, polymers are long chains of repeating units arranged in a totally random fashion. A good analogy would be a bowl of cooked spaghetti in a totally jumbled state—with a difference: unlike the cooked spaghetti, the polymers are in a constant state of motion. An even better analogy would be a bowl full of earthworms sliding against each other (yuk!). A thermoplastic where there is primarily only a random orientation of chains is known as an amorphous plastic.

Now imagine this bowl full of cooked spaghetti strands interspersed with neatly organized bundles of uncooked ones. This other type of structure with neatly bundled chains (called crystalline areas) mixed in with the random chains is called semicrystalline thermoplastics. See Figure 2.1.

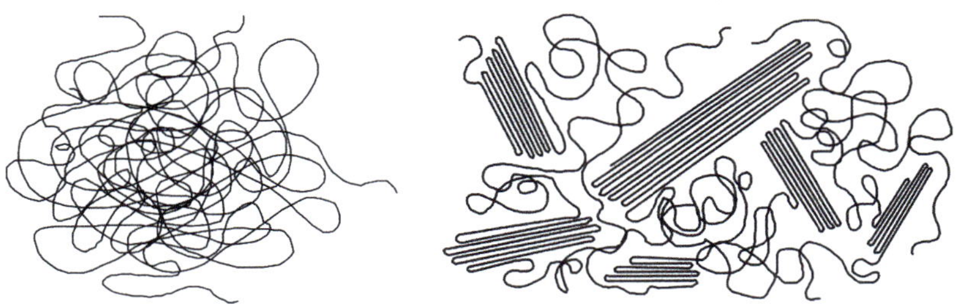

Figure 2.1 Amorphous and semicrystalline polymers

Common examples of amorphous plastics are polycarbonates (PC), acrylic–polymethyl methacrylate (PMMA), polystyrenes (PS), etc. The main characteristics of this family are that they:

- have no fixed melting point;
- have a range of temperature where the chains start to slide—a further increase in temperature can facilitate plastic flow;
- in many instances can be transparent;

- generally have lower tensile strength and higher impact strength (lots of exceptions!);
- have much lower and very predictable shrinkage;
- have uniform shrinkage in all directions.

All materials, however, are amorphous at injection, since the crystalline areas open up during the heating (more discussion on this in Section 2.2.6.2). One exception are LCPs–Liquid Crystal Polymers–which form areas of highly ordered structures even in the liquid phase, although the degree of order is less than that of a solid crystal.

Common examples of semicrystalline plastics are nylons, polyethylene, polypropylenes, etc. The main characteristics of this family are that they:

- have a distinct melting point: this is the point where the crystalline areas open up to lead to an amorphous state;
- generally are not transparent;
- have much higher shrinkage;
- have different shrinkage in flow and cross flow direction.

Again, all materials are **amorphous** at injection.

2.2.3.2 Thermosets

This book concentrates mainly on thermoplastic injection molded products. Therefore, we will just take a fleeting glance at thermoset materials. Due to the limited choice in the material types and molding processes, their use has been limited to a very few specific types of products requiring special properties (automotive tires being a *ubiquitous* example).

LIM (or Liquid Injection Molding) is used extensively for silicone parts. Another area that is extensively used is compression molded silicone parts, such as the one used in most keypads on handheld electronic devices–for instance, the TV remote controls today.

The process consists of two components either premixed or introduced as two separate components, as in the case of LIM. The premixed resins such as melamine formaldehyde–used extensively for inexpensive dinnerware and counter tops (melamine)–start with a mixture of melamine formaldehyde that is cross-linked (to be explained a little later) with alkyd, epoxy, and polyester resins, most generally in a compression molding process.

In LIM the two components (part A and B) are in liquid form. They are injected into the mold to react each other and get cross-linked.

Thermally, thermoset molding is the reverse of thermoplastic molding. In the former, cold material is introduced into a hot mold to react and cross-link. In the latter, hot material is injected into a relatively cold mold to cool down into the desired shape. As explained before, the former is a chemical process and cannot be reversed. The latter is a physical process and can be reversed by heating the plastic again.

Cross-Linking

Looking at Figure 2.2, the two side by side polymer chains have primary or covalent bonds in the length direction and secondary bonds (intermolecular forces, van der Waals, etc.) in the cross direction. The disassociation energy of the covalent bonds can be hundreds of times more than the secondary bonds. Thus, by supplying heat the secondary bonds are overcome and the chains can be made to slide against each other by applying pressure and generating the flow.

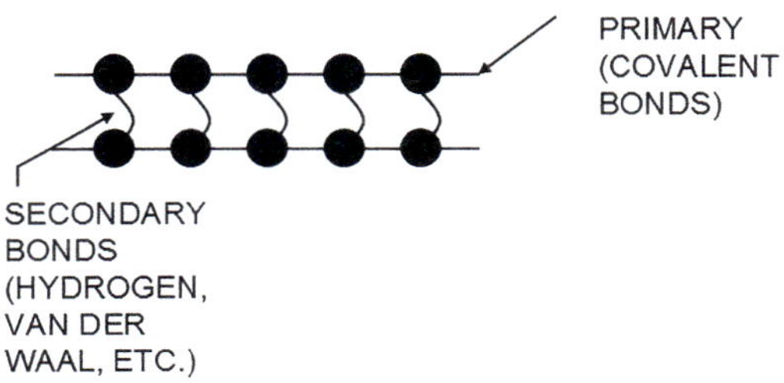

Figure 2.2 Primary and secondary bonds

Now let us look at Figure 2.3. In thermosetting resins, in the presence of a cross-linking agent and the application of heat, covalent bonds form between side branches also. Now the molecules are surrounded by covalent bonds in all directions, making it difficult to dissociate when heated again. The supplied energy is not enough to break all the bonds.

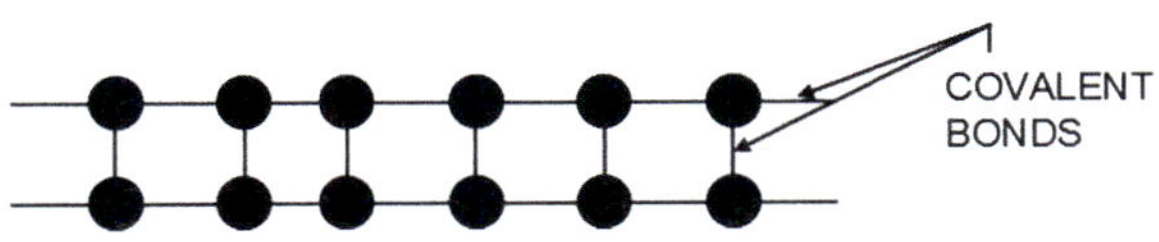

Figure 2.3 Cross-linked polymer

In one of the most common thermoset processes, sulfur is used to cross-link the sticky natural rubber chains into the high temperature and wear resistant tires that keep us safe at high speeds of travel.

One parting thought before we leave the area of thermosets. In the coming pages we will discuss the average molecular weight (AMW) of resins and its effect on their properties. As an example, polycarbonate, one of the toughest commercial plastics available, can have an AMW of 50,000. A thermoset polyurethane bowling ball, on the other hand, can be just one molecule–with the MW of 10^{28}!

2.2.4 Thermoplastic Molecular Arrangements

Thermoplastics can have different repeating structures. The most common ones are random copolymers, block copolymers, alternate, and homopolymers. Figure 2.4 illustrates these.

RANDOM COPOLYMERS

A-B-B-A-A-B-B-B-A-B-B-A

BLOCK COPOLYMERS

A-A-A-A-A-B-B-B-B-B-A-A-A-A-A

HOMOPOLYMER

A-A-A-A-A-A

ALTERNATING POLYMERS

A-B-A-B-A-B-A-B-A-B-A-B-A-B-A

Figure 2.4 Common polymer arrangements

In random polymers, two or more repeating mers form a chain in a totally random fashion. Block copolymers have organized blocks of the different repeating units forming the chains. The homopolymer has just one repeating mer in every chain. Finally, the alternating polymer has alternating mers as the name suggests.

By introducing and arranging these chains in these different fashions, properties can be altered and tailored to the requirements of the product. As an example, Table 2.1 compares the properties of two different types of acetals–a material commonly used in high wear and fatigue requirements, such as plastic springs and gears.

Table 2.1 Acetal Homopolymer and Copolymer

	HOMOPOLYMER	COPOLYMER
INITIAL MECHANICAL PROPERTIES	SLIGHTLY HIGHER	SLIGHTLY LOWER
PROCESSING TEMPERATURE RANGE	NARROWER	BROADER
MOLD DEPOSIT PROPENSITY	HIGHER	LOWER
EVOLUTION OF FORMALDEHYDE DURING PROCESSING	HIGHER	LOWER
STABILITY TOWARDS HEAT AGING	LOWER	HIGHER
ALKALI RESISTANCE	NOT RESISTANT	RESISTANT
RESISTANCE TO HOT WATER	LOWER	HIGHER

2.2.4.1 Blends, Alloys, and Copolymers

Most engineering plastics contain a variety of chains. The main types are blends and alloys.

Blends are a purely random mixture of two different polymer chains next to each other with no chemical bonds between them. A common example is PC/ABS—the blend of PC (polycarbonate) and ABS (Acrylonitrile-Butadiene-Styrene). The combination offers a lower price, slightly higher stiffness, and somewhat lower impact properties than polycarbonate.

In a copolymer, the different monomers are part of a common chemically bonded chain also referred to as the backbone. A common example of a copolymer is a PC-Siloxane Copolymer EXL from SABIC (formerly GE Plastics). This is a copolymer that builds on mechanical properties of polycarbonate by incorporating silicone into its molecular backbone structure. This process produces improved impact resistance and UV weatherability. It also improves the cold-temperature ductility to −40 °C, which allows a product to withstand prolonged outdoor exposure.

2.2.4.2 Commercial Plastics—Not Just Polymers

Most plastic polymers would be difficult to mold and would have little practical use if the resins consisted of just polymer chains. As discussed before, although polymer chains make up most of the plastic resins, there are many additional constituents that are needed to make the plastics moldable and have the long-term and/or the special properties that the parts made from them need. Some, but not all, of the constituents are:

- Antioxidants and heat stabilizers: to increase the heat and oxidation resistance needed for processing and offset weathering effects
- Colorants: for various colors

- Foaming agents: to create a foamed core for molding thick structural sections
- Plasticizers: to provide flexibility, such as flexible PVC for insulation
- Lubricants: to facilitate flow during molding and wear properties of the finished product
- Anti-stats: to reduce static electricity attraction
- Antimicrobials: to retard the growth of bacteria
- Flame retardants: to reduce flammability
- Fillers such as glass, carbon fiber, talc: to improve stiffness and other mechanical and/or electrical properties and to reduce cost, as in the case of talc.

Although the additives may improve certain molding or physical properties of the plastic, they are essentially "contaminants" to the basic plastic. A careful selection needs to be made in the types and amounts of these additives to make sure the molding or physical properties are not degraded in the process. As an example, a flame retardant may reduce the flammability but may severely degrade the molding and physical properties of the plastic.

2.2.5 Average Molecular Weight of Plastic Resins

One of the most important characteristics to control in the manufacturing and processing of the plastics is the average length of the chains and their distribution. Let us start with the basics.

2.2.5.1 Molecular Weight

Let us take you back to your high school chemistry class to start with the basics.

Atomic weight is the average mass of all the protons and neutrons in an atom. Figure 2.5 shows the carbon atom, which consists of 6 protons and 6 neutrons for an atomic weight of 6 + 6 = 12.

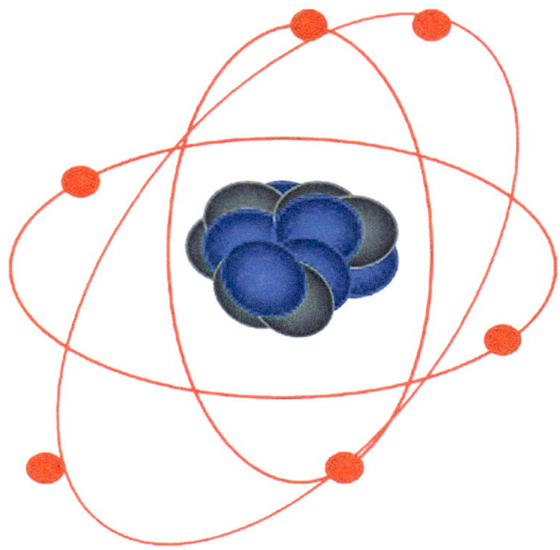

Figure 2.5 The carbon atom

The molecular weight is calculated as the sum of the atomic weight of each atom present multiplied by the number of atoms present in that molecule. As an example, hydrogen has an atomic weight of 1. Oxygen has an atomic weight of 16. Therefore, the molecular weight of water (H_2O) is $2 \times 1 + 16 = 18$. Refer to Figure 2.6.

Figure 2.6 The water molecule–H_2O

As stated before, a polymer is a repeating chain of many repeating units (monomers). Figure 2.7 shows a polystyrene monomer, which can also be written as C_8H_8. Its molecular weight, then, is $(12 \times 8) + (1 \times 8) = 104$.

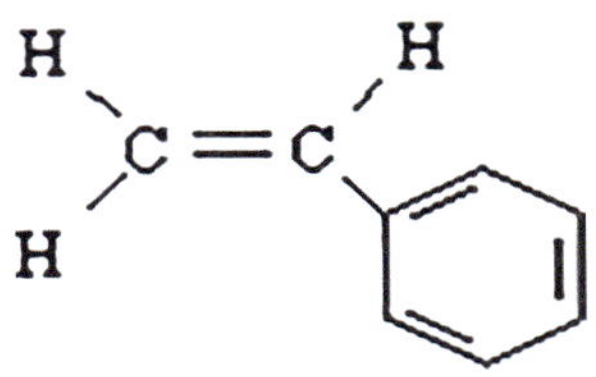

Figure 2.7 The styrene monomer

Figure 2.8 shows chains of 3, 5, and 7 styrene monomers. Their molecular weights are 3 × 104 = 312, 5 × 104 = 520, and 7 × 104 = 728, respectively.

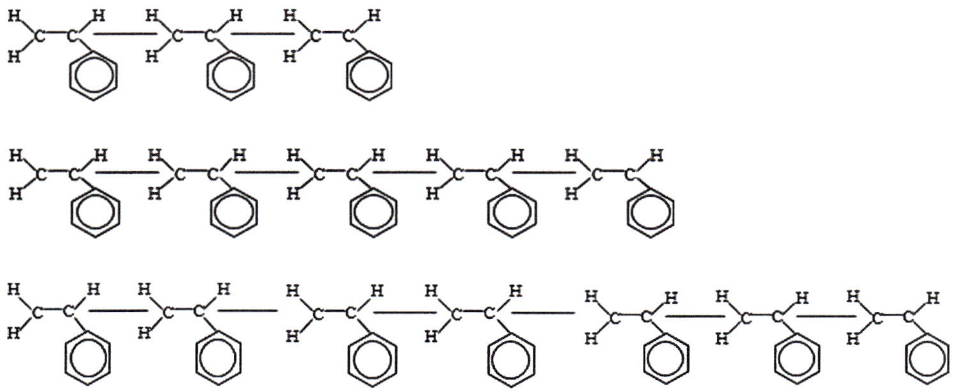

Figure 2.8 Polystyrene chains

2.2.5.2 Molecular Weight and Mechanical Properties

The AMW of the polymers is controlled very precisely in the manufacturing of the resin to make sure the distribution is very "tight." Figure 2.9 shows the distributions for the same polymer for two different AMW. The one with the higher AMW is stronger than the one with the lower AMW. The reason is that the former has much longer chains–leading to a higher degree of entanglement–than the latter, as shown in Figure 2.10.

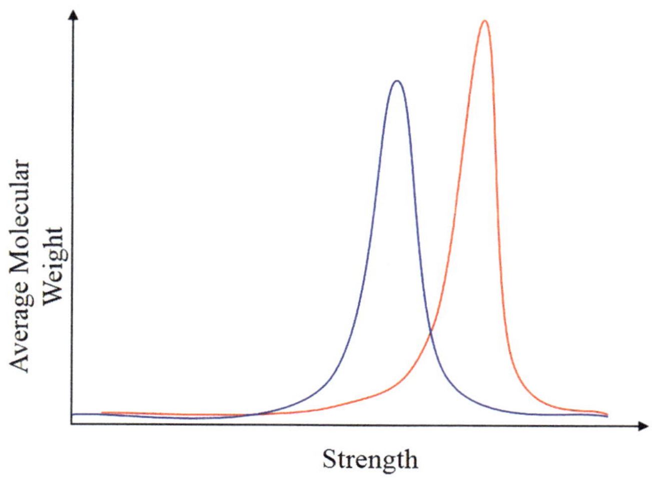

Figure 2.9 Average molecular weights and distribution

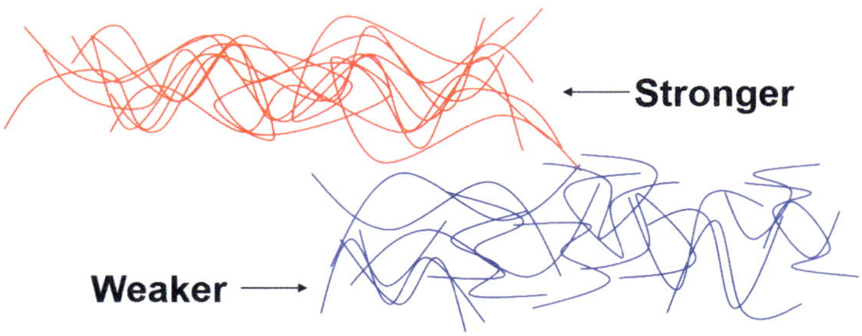

Figure 2.10 The greater the chain length, the greater the entanglement

Try this for yourself. Take a roll of strings and cut them into two different bunches. Cut the strings so that the length of the first bunch is about twice the length of the second bunch. Now jumble both bunches individually, keeping them separate from each other. Holding the two ends of each bunch, try to pull the strings apart. You should notice that there is a measurable difference between the force required to separate the longer strings bunch versus the shorter strings one.

Candle wax has the repeating monomer shown in Figure 2.11.

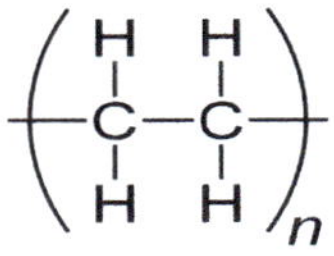

Figure 2.11 Candle wax–ethylene monomer

Polyethylene resin is nothing but longer chains of the same. Below are the AMWs for the various forms of "wax."

- Candle wax ~ 200-600
- Low to high density polyethylene (LDPE, HDPE) (milk jugs) ~ 10,000
- Ultra-high molecular weight polyethylene (UHMWPE) (competes with aramid[1] in bullet proof vests) ~ 3.1 to 5-67 million

2.2.5.3 Molecular Weight and Viscosity

Back to high school physics: Viscosity is the property of a fluid that resists the force causing the fluid to flow. The higher the resistance, the higher the viscosity. Wax, honey, and water have the highest to lowest viscosity, respectively.

[1] Aramid or aromatic polyamide fibers are a type of heat-resistant and strong synthetic fibers used in aerospace and military applications, body armor fabric, and ballistic composites. Kevlar® is one of these from Dupont.

The AMW is directly related to the viscosity. The higher the molecular weight, the higher the entanglement–hence viscosity–and, as discussed before, the higher the strength.

2.2.5.4 Melt Flow Rate

We will now introduce and discuss the term melt flow rate (MFR), also known as melt flow index (MFI). The ASTM (American Society for Testing and Materials–an international standards organization that develops and publishes technical standards for a wide range of materials, products, systems, and services) describes the MFR as the following:

ASTM D 1238

The flow in grams of a material in 10 minutes through an orifice 0.0825 inches in diameter under a certain force and a certain temperature.

In the case of polycarbonate the force is 1.2 kg and the suggested temperature is 300 °C.

Figure 2.12 shows a typical MFR tester.

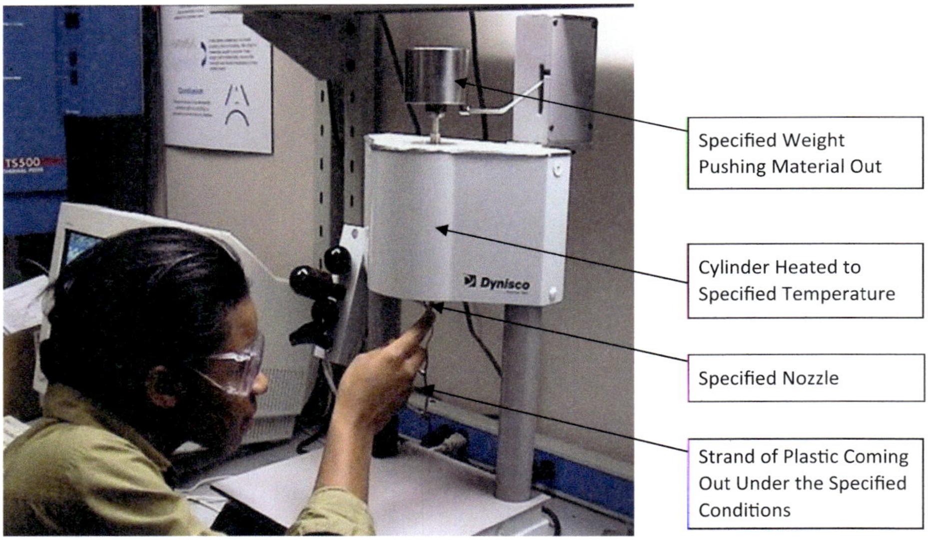

Figure 2.12 MFR tester

2.2.6 Injection Molding of Thermoplastic Materials

2.2.6.1 Heating Amorphous and Semicrystalline Materials

As the material is heated, the chains loosen up. Heating to a point above the processing range can cause the degradation of the polymer and the additives.

Figure 2.13 shows the heating of amorphous resins and Figure 2.14 that of semicrystalline materials. The main difference is that the former does not have a distinct melting point. In its place, it has a range of temperature where it can be injection molded under sufficient pressure. Semicrystalline materials, on the other hand, have a distinct melting point, T_M, where the crystalline areas open up.

The T_G in the figures stands for glass transition temperature, which is defined as the transition in amorphous materials (or in amorphous regions within semicrystalline materials) from a hard and relatively brittle state into a rubber-like state. T_G is more distinct in amorphous materials. This change is reversible.

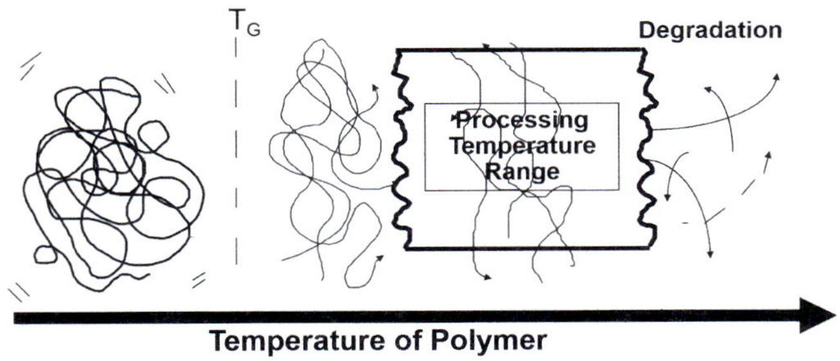

Figure 2.13 Heating of amorphous materials [1]

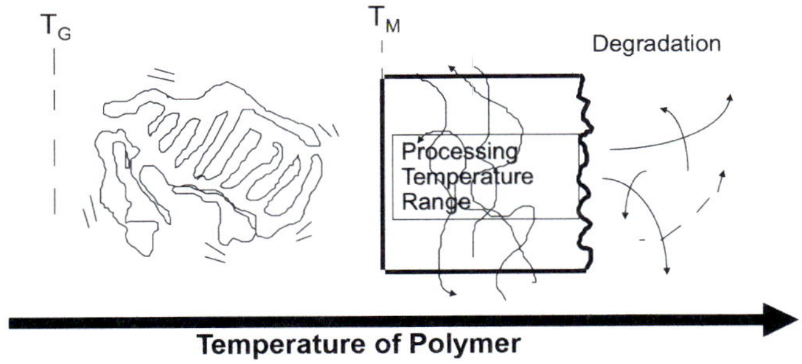

Figure 2.14 Heating of semicrystalline materials [1]

2.2.6.2 All Materials Are Amorphous During Injection

Earlier it was stated that amorphous materials can be transparent. It was also stated that all materials become amorphous when heated to the right temperature. Figure 2.15 shows what happens to wax (a semicrystalline material) when it is heated. In going from left to right the candle is heated from room temperature to beyond its melting point. One can see that the crystalline areas have completely opened, rendering the wax totally transparent.

Figure 2.15 Wax turning amorphous and therefore transparent when heated

2.2.6.3 Degradation

Though degradation occurs very fast at higher temperatures, given enough time, degradation can occur even at lower temperatures. Thus, materials left in the heated barrel of an injection molding machine after the heats have been turned off continue to degrade over time and need to be purged before starting production again. Too long a residence time in an oversized barrel can degrade the material also in normal molding. More discussion on this can be found in the chapter on processing (Chapter 5).

Degradation also needs to be taken into account when plastic parts are designed to be used under continuously elevated temperature.

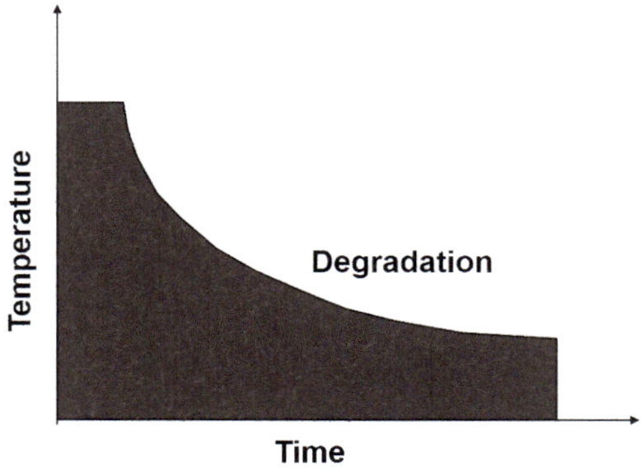

Figure 2.16 Degradation as a function of time and temperature

2.2.7 Plastic Properties

Having come from the metals world, most of us consider just the properties that we would for a conventional metal part design. As I will propose again and again, plastics are not metals and therefore we need to look at a whole set of properties that we did not have to for metals. Even the conventional properties, such as tensile strength, modulus of elasticity, etc., have to be used with caution. Below is a list of properties that need to be considered; this list will be explained in detail in various parts of the book:

- Basic physical properties
  - Static: mechanical–tensile strength, toughness, elongation, etc.
  - Dynamic: wear, modulus at high impact rate, fatigue
  - Electrical: dielectric strength, resistance, etc.
  - Long term: color retention, embrittlement
- Environmental properties
  - Chemical resistance
  - Color and appearance
- Thermal properties
  - High and low temperatures
  - Temperature cycling
- Agency requirements
  - UL
  - FDA
  - RoHS[2]
  - REACH[2]
  - Country specific requirements
- Molding properties
  - Flow: MFI, flow length vs. thickness
  - Bonding between materials (overmolding)
  - Directional shrinkage

2.2.7.1 A Word of Caution about Using MFI to Predict Flow

It is a common mistake to use the MFR or the MFI index to compare the flow of two different materials during molding. As discussed before, we assume that a higher MFR means a higher flow and therefore a lower viscosity. The issue is that the MFI

[2] More information in Chapter 9.

is measured at very low shear rates and molding occurs at very high ones. More details follow.

Viscosity, Shear Rate, and Newtonian Flow

At this point we need to have a discussion on Newtonian and non-Newtonian fluids, shear rate, and their relationship with viscosity. See Figure 2.17.

A Newtonian fluid is one whose viscosity is independent of the shear rate. Shear rate is the rate at which the liquid is being discharged or pumped. Water falls into this category. If you are pumping water through a hose, its viscosity remains the same whether it is just dripping or being pumped in buckets per second.

A non-Newtonian fluid, on the other hand, will not flow at all or flow very slowly if it is being pumped at a very slow rate. However, if the pumping rate is increased, its viscosity will drop inversely proportional to the rate.

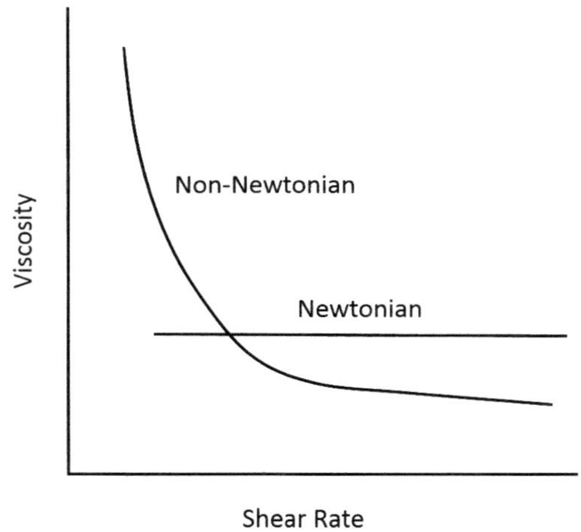

Figure 2.17 Newtonian and non-Newtonian flow

The MFI is measured at an extremely low shear rate compared to the injection molding process, where the shear rate may go up to 40,000 or more reciprocal seconds (1 divided by the number of seconds to pump a specific volume). Now look at Figure 2.18. The MFI is measured at a very slow rate (60 seconds to pump a few cubic centimeters of plastic). At this rate the viscosity of MFI 8 is higher than MFI 16. At the shear rate encountered in injection molding (40,000 reciprocal seconds or more) the MFI 8 flows much more easily than MFI 16!

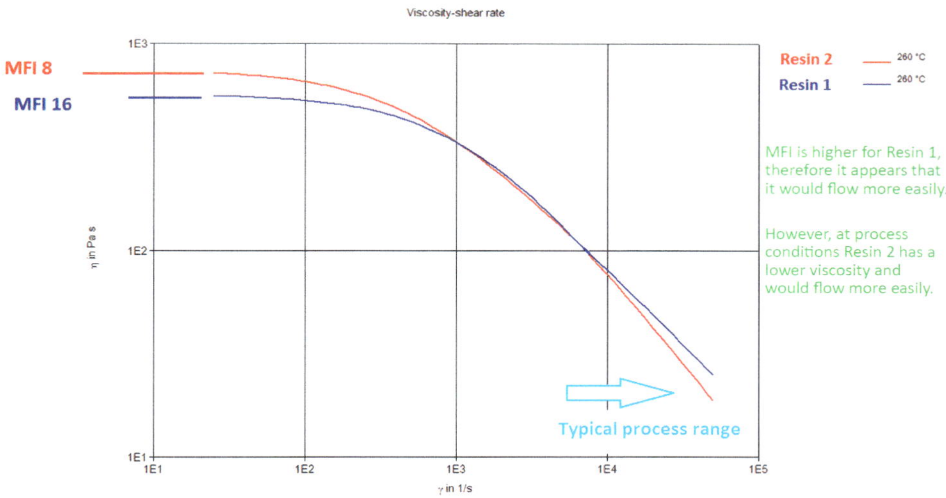

Figure 2.18 MFI vs. viscosity at differing shear rates [11]

2.2.8 Color Fundamentals

As will be discussed in Chapter 8, a big portion of plastic parts are rejected because of visual issues. Color mismatch is a big part of the visual defects. In order to make the job of color specification, measurement, and control a little easier let us review some color fundamentals.

The color that one perceives is based on the light source, the pigment or dye, and the human eye looking at the colored object. See Figure 2.19.

Figure 2.19 Color perception

2.2.8.1 Color Measurement

There has been a lot of work done on the measurement of color started by CIE–Commission Internationale de l'Éclairage (International Commission on Illumination). In order to specify a color, two sets of values are needed. The first is the identification of the particular color in the entire color space. The second is the gloss level (shiny or dull).

2.2.8.1.1 Color

Color is specified in terms of hue, value, and chroma. Figure 2.20 shows this. Looking at Figure 2.20:

- Hue describes the family of the color–namely, green, red-yellow, etc.
- Value indicates the lightness of the color from white to black
- Chroma describes the saturation of the color

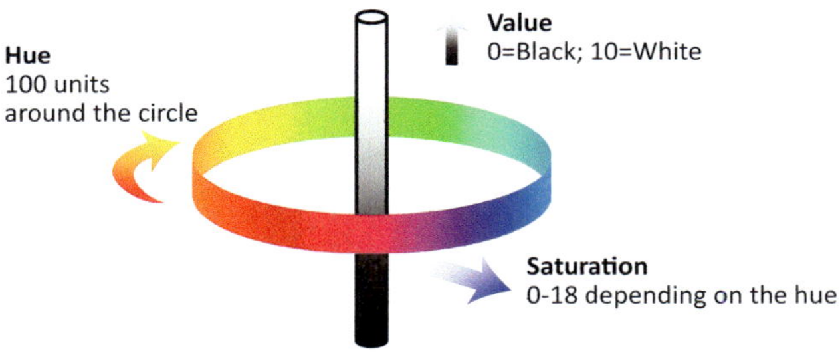

Figure 2.20 The color space–hue, value, and chroma [5]

The most common system of measurement of color today is the Hunter L, a, b system. See Figure 2.21.

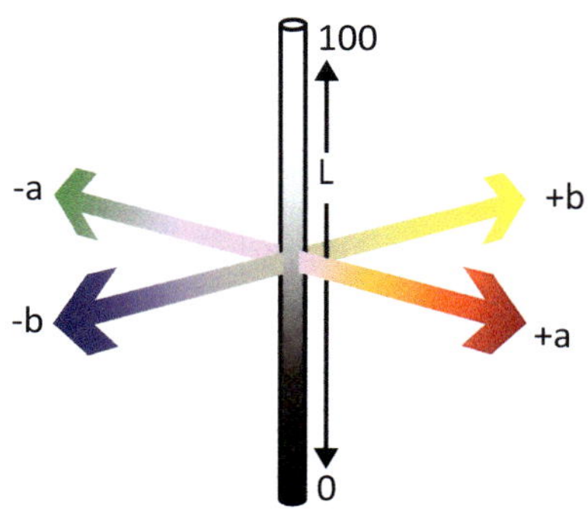

Figure 2.21 The Hunter L, a, b color measurement system [5]

The Hunter L, a, b color space is organized in a cube form. The L axis runs from top to bottom. The maximum for L is a hundred, which would be a perfect white color. The minimum for L is zero, which would be pure black. The a and b axes have no specific numerical values. Positive a is red. Negative a is green. Positive b is yellow. Negative b is blue.

There are two popular L, a, b color scales in use today. Hunter L, a, b (1958) and CIE L*, a*, b* (1976). The current CIE recommendation is to use the latter.

The color difference between the master and the product color is called delta E or ΔE. The mathematical formula for ΔE is:

$$\Delta E = \sqrt{\Delta L^2 + \Delta a^2 + \Delta b^2} \tag{2.1}$$

Most plastic products are allowed to have a ΔE of 1, which in theory is supposed to be an appearance difference that an average human eye cannot detect. In reality, depending on the criticality of the requirement, the maximum ΔE permissible may vary from a fraction to more than 1.

ΔEs are three-dimensional spaces generally oval in shape. They can be different for different locations in the color space as seen in Figure 2.22. David MacAdam [3] conducted experiments with actual observers and came up with these ovals named as MacAdam units of color difference.

As can be seen in Figure 2.22, in the light green area the ellipses are much bigger than the ones in the blue area. Thus the green area can have a bigger color difference than the blue area and still be undetectable.

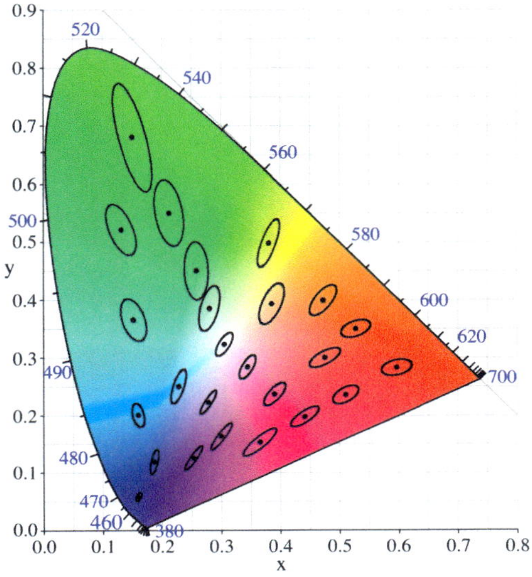

Figure 2.22 MacAdam ellipse [4]

Additionally, some color difference attributes are more objectionable than others, as indicated below–highest to lowest:

- Hue
- Chroma
- Lightness or darkness

ΔE is not always reliable by itself. In the example shown in Figure 2.23, batch 1 is visually a good match to the standard. Batch 2 is not. However, they both have the same ΔE value. For batch 2, all the difference is in the "a" value (less green) and it is visually unsuitable.

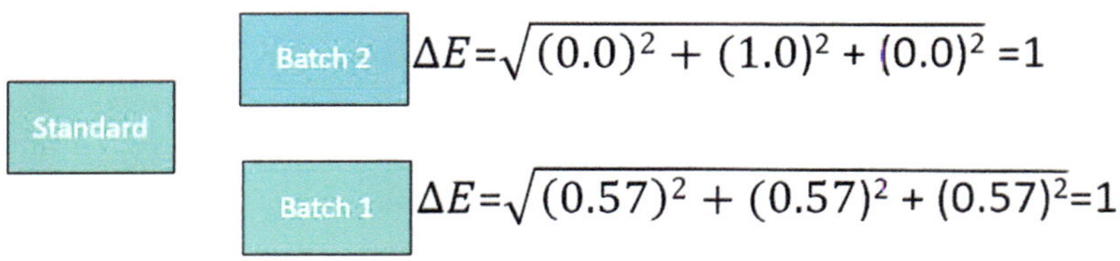

Figure 2.23 Color difference example

To conclude the above discussion, my suggestion is not to specify the acceptable color difference as a one size fits all ΔE. Rather increase or decrease it based on the application and the location of the hue in the color space. It can further be refined as an overall ΔE with specific limits on ΔL, Δa, and Δb.

Metamerism

Before we leave the topic of color matching, let us look at another "gotcha" lurking in the bushes. We mentioned earlier that the appearance of a color depends on the observer, the pigment, and the light. As it turns out, certain colors may or not match depending on certain combinations of light, observer, and the color and geometry of the pigments. This is called "metamerism."

The most offensive of these is the color–with different "a" and "b" values. The color may appear same or different depending on the light in which they are seen. In most plastic applications this is the metamerism to be avoided.

Figure 2.24 shows the visible spectrum and the corresponding wavelengths in nm. From left to right, the colors are violet, indigo, blue, green, yellow, orange, and red (VIBGYOR), with wavelengths going from 400 to 700 nm.

Different light sources emit different frequencies. Figure 2.25 shows the spectrum generated by daylight, incandescent light, and fluorescent light.

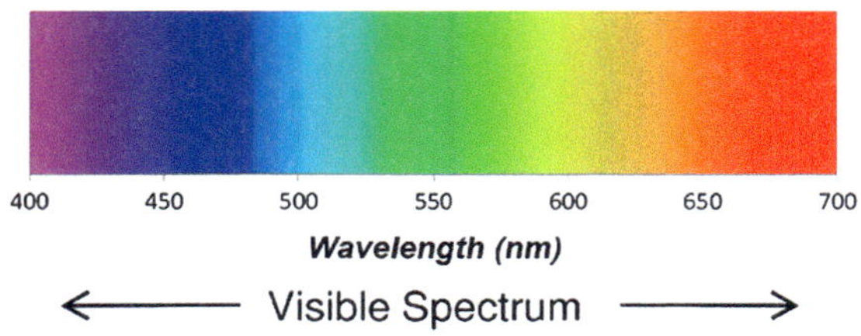

Figure 2.24 The visible spectrum

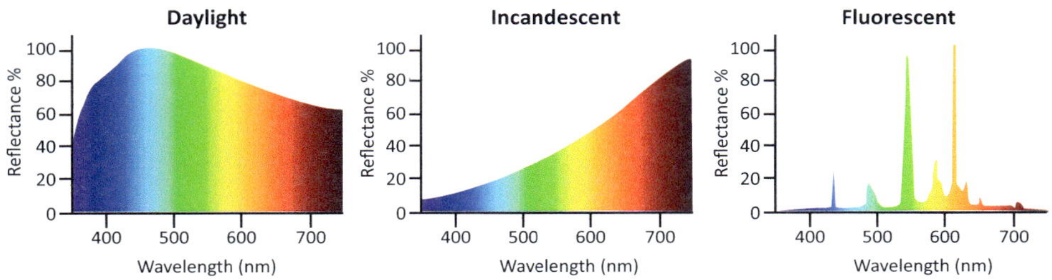

Figure 2.25 Common light sources

Figure 2.26 is called a spectrophotometric or reflectance vs. light frequency curve. Ideally, the curves for the standard and the match should totally match each other. In this case, the wavelength vs. reflectance curves indicate a metameric pair. The Y axis indicates the reflection and the X axis represents the color of the light source. The orange curve represents the standard and the purple the matched plastic part. Looking at Figure 2.25 and Figure 2.26, in the daylight spectrum, the colors match and look orange red. However, in the predominant peaks of green and orange red, the color looks more purple.

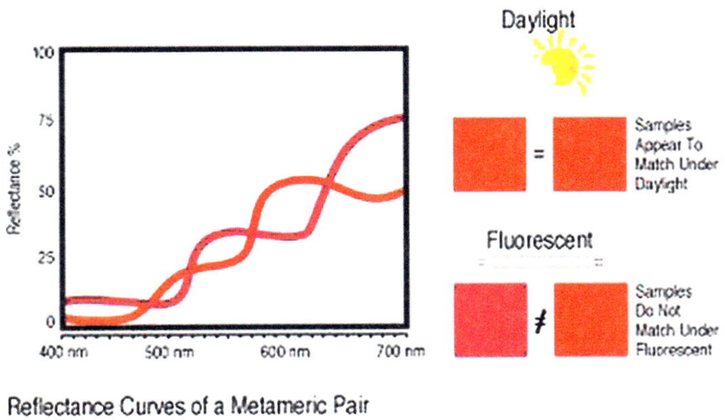

Figure 2.26 Reflectance curves of a metameric pair [1]

To avoid this, spectrophotometric curves should be obtained along with the L, a, b, and ΔE values when matching a color. The curves for the match and the standard should closely resemble each other without crisscrossing.

2.2.8.1.2 Gloss

The second component in the appearance of a color is gloss. Gloss affects the appearance in two ways:

The obvious–how shiny or dull a surface is. In the case of plastic parts, this can depend on:

- The polish of the steel surface producing the part
- The material itself–polycarbonate is inherently glossier than the polycarbonate/ ABS (PC/ABS) blend
- Long term exposure of the part to various lights and the consequential dulling

Gloss Level Measurement

Looking at Figure 2.27, the gloss level is simply the percentage of light reflected by a surface of the incident light. The unit of measure is gloss level unit (GLU).

There are three types of instruments based on the angles of incidence and reflectance of the light. The standard meters have angles of 85°, 60°, and 20° for measuring from very dull to very shiny surfaces, respectively. ASTM D 523 specifies that 60° can be used for most materials. This is the instrument used to measure most plastic parts.

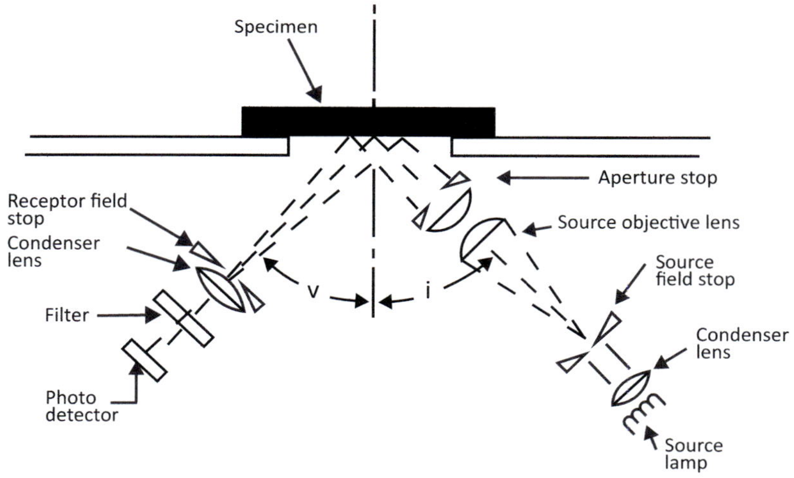

Figure 2.27 Gloss level measurement

The second aspect of the surface finish is the naturally reflected light. About 4% of the incident normal light (generally white) is reflected back directly from the surface appearing as white light. The rest is refracted by the pigment under the surface. A dull surface has a lot more surface area than a shiny surface. See Figure 2.28. Therefore, a lot more white light is reflected from it making it appear lighter than the shiny surface.

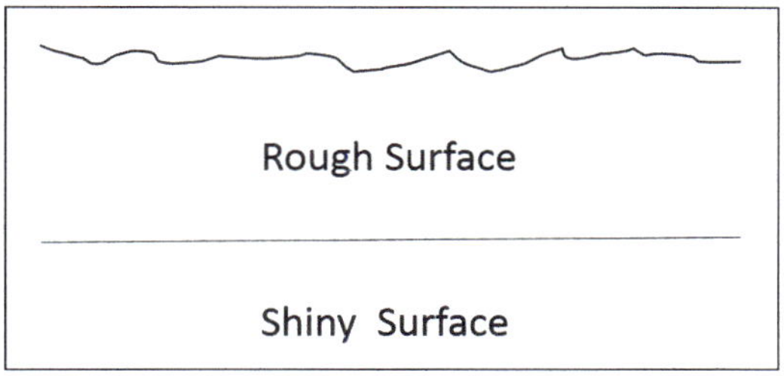

Figure 2.28 Reflective areas of shiny and dull surfaces

Figure 2.29 shows a polished area (left) and one textured with Mold-Tech® texture MT 1010 on a **common** injection molded plate in ABS. The textured area (with a consequential higher surface area) appears much whiter than the adjacent shiny area.

Figure 2.29 Dull and shiny surfaces side by side

2.3 "Plastics Are Not Metals"

2.3.1 Color, Gloss, and Aesthetics

Now that we have covered the basics of polymeric materials, let us look at how they are different from metals. Most of the engineers and designers come from the metals world. Therefore, most of us make assumptions on the predicted performance of plastic properties based on our metals background.

I started with a discussion on color and appearance because this knowledge is extremely important in the case of plastics. Most plastic parts have dual functions—aesthetics and physical performance. This is due to the fact that very few of them need to be painted or otherwise decorated if designed and manufactured with due diligence.

On the other hand, even if we are designing the most aesthetically critical metal components such as exterior automotive parts, we mostly choose the metals and alloys based on the physical properties, weight, and cost. The aesthetics are left to the paint specialist, who will in most cases find a paint system (primer, paint, and application method) that will meet the cost, durability, and the cosmetic requirements. In other words, aesthetics and physical properties are quite independent of each other. A vast majority of metal parts meet their aesthetic and environmental requirements just by getting plated, getting chromate conversion coated, or being anodized.

Plastic parts not only need to meet the short term color and appearance requirements, they also need to be resistant to long term color shift and fading.

In Figure 2.30, the first set of pictures is of air conditioning ABS vents in my own house. Some of them turned from white to various shades of yellow to brown in a matter of months when exposed to the normal sunlight entering the house. The others retained the original white color.

Both covers for the TV are made of ABS. One remained the original white and the other turned yellow.

The middle picture is one of my neighbor's PVC mailbox, a few months after installation. In this case the plasticizer from the PVC completely evaporated, leaving the mailbox very dull and also brittle. (For those old enough, remember the cracks across the speaker grills in the PVC dashboards of the cars of the seventies and eighties?)

The bottom picture shows a very similar mailbox before exposure to the environment.

The long term properties of plastic resins are carefully modified by adding UV stabilizers and antioxidants by the resin suppliers. One just needs to choose the resin with the right protection.

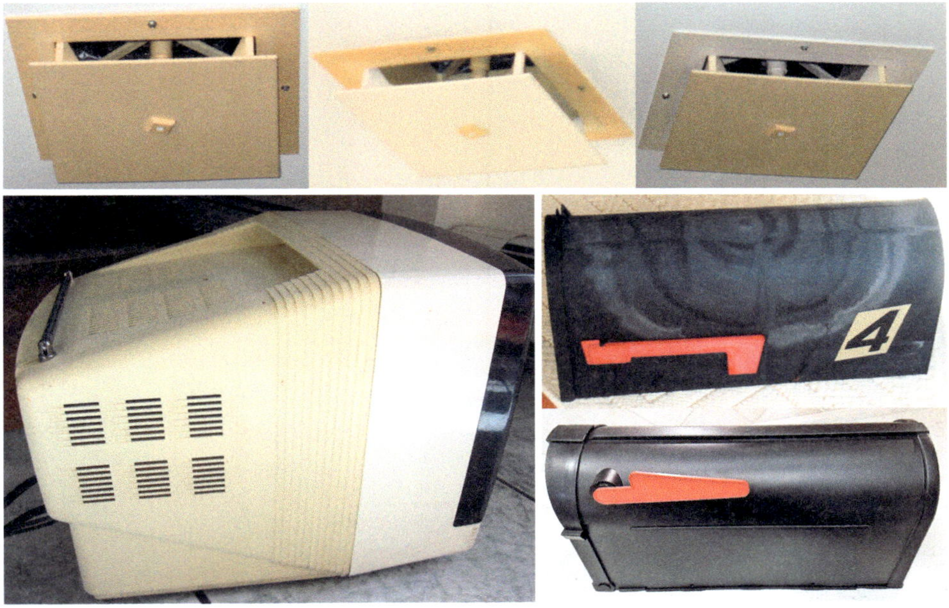

Figure 2.30 Long term yellowing and fading of plastics

Figure 2.31 shows the yellowing and fading of various GE (now SABIC) plastics. Figure 2.32 and Figure 2.33 show that fading and color shift are very dependent on the colors even in the same resin and that care should be taken in the color selection. In the case of color retention, green is the worst and bright white and black are the best.

The same applies to gloss retention. Here the green performed the best but the white and black did not do too badly.

Notice that Delta b (dB in Figure 2.31) is used for the yellowness shift and Delta E (dE in Figure 2.32) for color. Going back to our previous discussion, an increase in the b value indicates that the color is more yellow. Delta E, obviously, indicates total color shift.

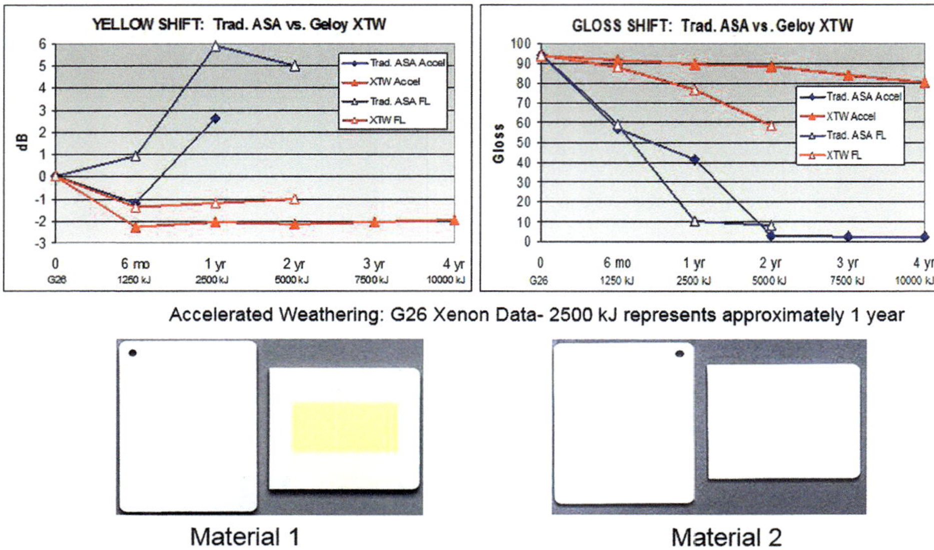

Figure 2.31 Yellowness and gloss shift in plastics[3] [1]

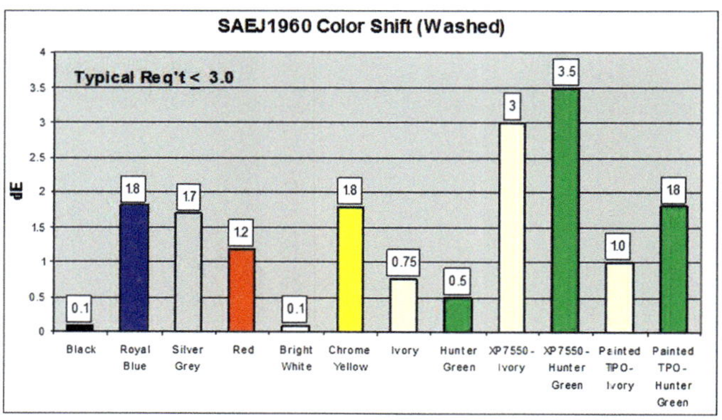

Figure 2.32 Color shift in plastics–SAE J1960 color shift[4] [1]

[3] GELOY is a trademark of SABIC or its subsidiaries or affiliates.

[4] SAE accelerated exposure of automotive exterior materials using a controlled irradiance water-cooled xenon arc apparatus.

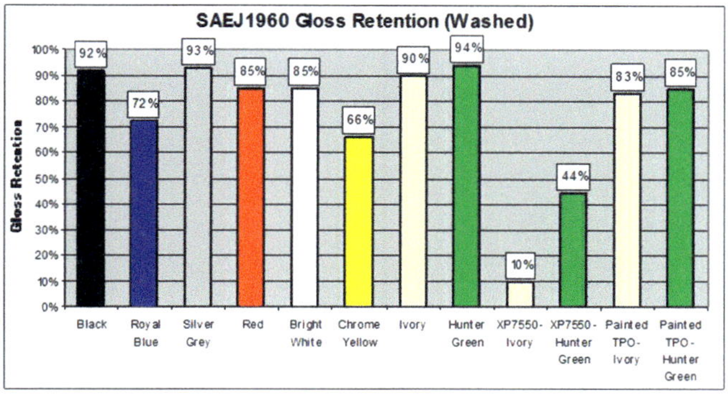

Figure 2.33 Gloss retention [1]

2.3.2 Long Term Stress

Plastics do not take long term stress well[5]. Let us juxtapose a plastic component with a metal component in a very simple example. Hang a mass of 100 kg from any structural metal rod **designed to withstand the induced stress**. Measure the starting length. Leave a note for your great-great-grandchildren to measure it a hundred years later. Will they find a change in the length when they measure it? Probably not. OK, maybe in the case of zinc, we may find a failure. Zinc and some other low melting point alloys do suffer from some creep phenomenon.

5 Image above: anonymous.

Do the same for a plastic cross section **that is designed to take the stress of the force of a hanging 100 kg mass**. You better not stand under the weight! The weight may drop on you within minutes, depending on the specific plastic and the environmental conditions. All plastics are viscoelastic and are subject to creep and stress relaxation. Under continuous stress their properties will change to the point of rupture.

2.3.3 Plastics and Chemical Resistance

Although both metals and plastics can degrade due to environmental exposure over time, the degradation of metals is a lot more predictable and controllable. The most common alloy and degrading phenomenon are steel and rusting. With the right paint or plating and with periodic maintenance, the rusting can be kept under control for centuries.

A great example of the longevity of metals is the iron pillar in front of the Qutub Minar in Delhi, India. Though the pillar was forged about 1600 years ago and moved to Delhi nearly 1000 years ago, it still stands fully intact and free of rust.

Figure 2.34 The iron pillar in front of the Qutub Minar in Delhi [12]

Also, the level of stress in the metal structures plays almost no role in how fast the metal will oxidize or be affected by common environmental chemicals such as acids, alkali, salt, or solvents. Even galvanic corrosion is very predictable and controllable by using the right pairs of metals in contact with each other.

Now let us step into the world of plastics.

Different plastics have different resistance to common solvents and chemical agents. Even moisture in the air can degrade plastics by breaking down the polymer chains. Plastics can also get oxidized and lose their mechanical properties.

Here is the key difference, though. Plastics under even a small amount of tensile stress are a lot more vulnerable to Environmental Stress Cracking (ESC) and chemical attack than those without it. Generally, amorphous plastics are more vulnerable than semicrystalline ones.

Let us look at the reason why.

As discussed previously, most polymers have long chains mostly of hydrocarbon molecules. When the plastic is cooled down from a high temperature, the chains re-entangle themselves. In the process they leave some spaces between the chains, which are called the "free volume." As a side note, because of the orderly structures of the crystalline areas in semicrystalline materials, the free volume is less and therefore they are more chemically resistant than their amorphous counterparts. Most oils and solvents are also hydrocarbons and have a natural affinity to the hydrocarbons in the polymer. If allowed to enter the free spaces, they can cause the chains to break. If the chains are packed tightly, the attacking agents have less opportunity to enter and cause the degradation [2].

Now let us look at the typical injection molded wall. See Figure 2.35.

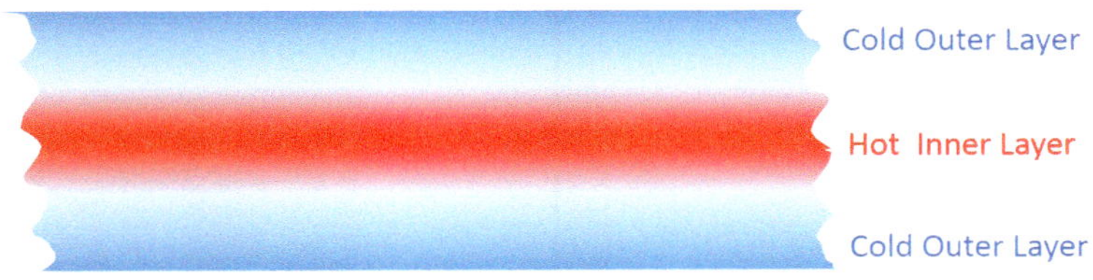

Figure 2.35 Cooling of injection molded wall

Plastic basically is a very poor conductor of heat. An injection molded part may be ejected out of a mold at upwards of 200 °C. As soon as it comes out, the outer layers are exposed to the ambient air and start to cool and shrink. The inside layers are insulated by the outside layers and stay hot for a much longer time.

The net result is that the outside layers are continuously pulled in by the inside layers and the inside layers are pulled out by the outside layers when the part has finally cooled down. This results in the stress pattern for the part shown in Figure 2.36.

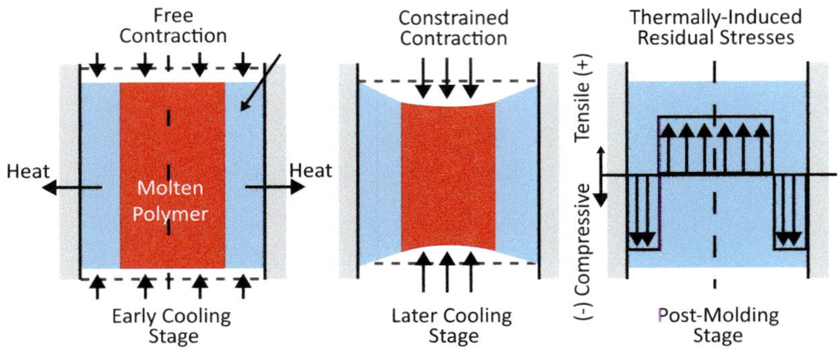

Figure 2.36 Stresses induced during cooling [6]

2.3.3.1 Physical Effect

To actually demonstrate the above stress pattern and the effect it has on its physical properties, the top surface of a 3 mm thick PC dog bone was machined about 0.5 mm deep. Care was taken to make sure the plastic stayed at room temperature during the machining, by circulating plenty of cold water at the interface of the cutting tool and the surface being machined. As soon as the piece was released from the holding fixture after machining, the dog bone curled upwards. The top layer of compressive stress was removed and tensile stress in the next layer pulled the two ends in. Whereas the original dog bone could be bent hundreds of times without a fracture, one bending cracked the part in the middle! See Figure 2.37.

As I will mention in Chapter 3, this is one reason I strongly advocate against using machined plastic parts for doing any kind of impact or physical testing during the development of a project.

Figure 2.37 Stress relief with machining

2.3.3.2 Chemical Resistance

Let us also look at what it does to the free spaces. In a compressive layer the free spaces are reduced and in the tensile layer the free spaces are increased. Therefore, as molded, any plastic part is naturally more resistant to ESC and chemical attack.

What happens when the part is made to bend (or "unbend" in the case of a warped part)?

Figure 2.38 shows the simulation of a dog bone being bent. As can be seen from these pictures, the compressive layer quickly turns into a tensile one.

The fact that the load carrying capacity of a plastic part is in a state of continuous stress may be secondary to the fact that it has now become more susceptible to chemical attack and ESC. Never mind strong chemicals, it can now be made to fail even with the humidity in the air.

In the training I conduct, I take a PC dog bone and flex it back and forth tens of times. I then smear it with acetone and leave it aside. I take another dog bone and bend it very slightly to induce a tensile stress on it. I put a small drop of acetone on it. It completely breaks apart into two pieces. See Figure 2.39. At the end of the session I look at the piece that I smeared with acetone and bend it back and forth. Practically no damage is done because the acetone evaporated before it could penetrate the surface with the compressive layer under it.

Please visit my website to view a video of the above. Here is the link: *www.vikpedia.org*.

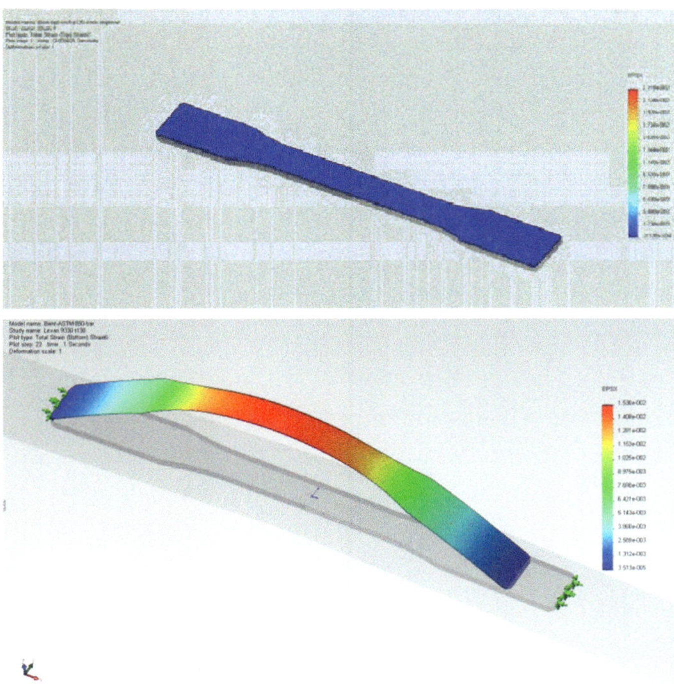

Figure 2.38 Stresses induced in bending [7]

Figure 2.39 Bent dog bone—catastrophic failure with one drop of acetone [11]

2.3.4 Physical Properties—Plastics vs. Metals

Table 2.2 shows the salient differences between plastics and metals.

Published Properties

Since the actual properties are so dependent on the way testing is performed as well as the actual conditions of use (time, temperature, rate of loading, environmental conditions, molding conditions, physical design, tooling, etc.), the published plastic properties are more of a set of guidelines, as a starting point, and for comparing different plastics. As will be explained in Chapter 3, these properties have to be used with a lot of caution. In fact, most resin suppliers have disclaimers to this effect to make sure the users do not misunderstand the properties and apply them incorrectly. See Figure 2.40. This is a typical disclaimer most resin suppliers use.

Michael Sepe is an industry leader as a material analyst. Some of his thoughts on the data sheets are that they:

- Provide short-term mechanical properties at room temperature
- Provide values for yield stress and impact energy to break and tell us when catastrophic failure occurs but nothing regarding what works
- Do not provide a clear picture for the effects of time under load
- Are only an attempt to deal with the effects of elevated temperature—DTUL (Deflection Temperature Under Load) and Vicat softening (which can also be very misleading)

I encourage the readers to read some of his work to get more insights on what these properties are and provide.

Table 2.2 Plastics vs. Metals Salient Physical Differences

PLASTICS	METALS
Published properties are only guidelines	Published properties are reliable for design
Properties affected by environment, temperature, time and are relatively unpredictable	Properties not affected by NORMAL environment, temperature, time. Even if they are, they are more readily predictable
Unique design requirements	Conventional design requirements
Tooling and processing can have dramatic effect on performance	Tooling and processing have relatively little effect on performance
Many latent defects that can not be detected with routine quality control	Few latent defects that can not be detected with routine quality control

THESE PROPERTY VALUES ARE NOT INTENDED FOR SPECIFICATIONS PURPOSES...

(1) Typical values only. Variation within normal tolerances are possible for various colors. All values are measured after at least 48 hours storage at 23 °C/50% relative humidity. All properties, except the melt volume and melt flow rates are measured on injection molded samples. All samples tested under ISO test standards are prepared according to ISO 294.

(2) Only typical values for selection purposes. ***Not to be used for part or tool design.***

(3) This rating is not intended to reflect hazards presented by this or any material under actual fire conditions.

(4) Internal measurements according to UL standards.

(5) Measurements made from laboratory test coupon. ***Actual shrinkage may vary outside of range due to differences in processing conditions, equipment, part geometry, and tool design.*** It is recommended that mold shrinkage studies be performed with surrogate or legacy tooling prior to cutting tools for new molded article.

.... Each user is responsible for making its own determination as to the suitability of the seller's products, services, or recommendations for the user's particular use through appropriate end-use testing and analyses.

....

Figure 2.40 Typical disclaimers

2.3.4.1 Temperature Dependent Behavior of Plastics vs. Metals

Refer to Figure 2.41.

Looking at PC, although the continuous use temperature stated is about 125 °C, one can see that the elastic modulus drops from over 2,500 MPa to about 2,200 MPa at less than 100 °C, or more than 10%. Thus if one designs a component with the properties at 23 °C, the performance may be very different at 100 °C. If the use of temperature goes up to 150 °C, which can happen in an isolated area in an electronic enclosure, the modulus literally falls off a precipice.

Looking at the metals for comparison, first of all we are looking at a value of 2,500 MPa vs. over 200 GPa in the case of a carbon steel–the difference of a few hundred MPa is in the "noise" category.

Compared to the published plastic properties, metal ones can be "taken to the bank." As an example, if the ultimate tensile strength of one of the most common alloys, Cold Rolled Steel (CRS - AISI 1020), is about 420 MPa or 61,000 psi, this alloy from ANY manufacturer will have a dependable similar property that the designer can use with confidence.

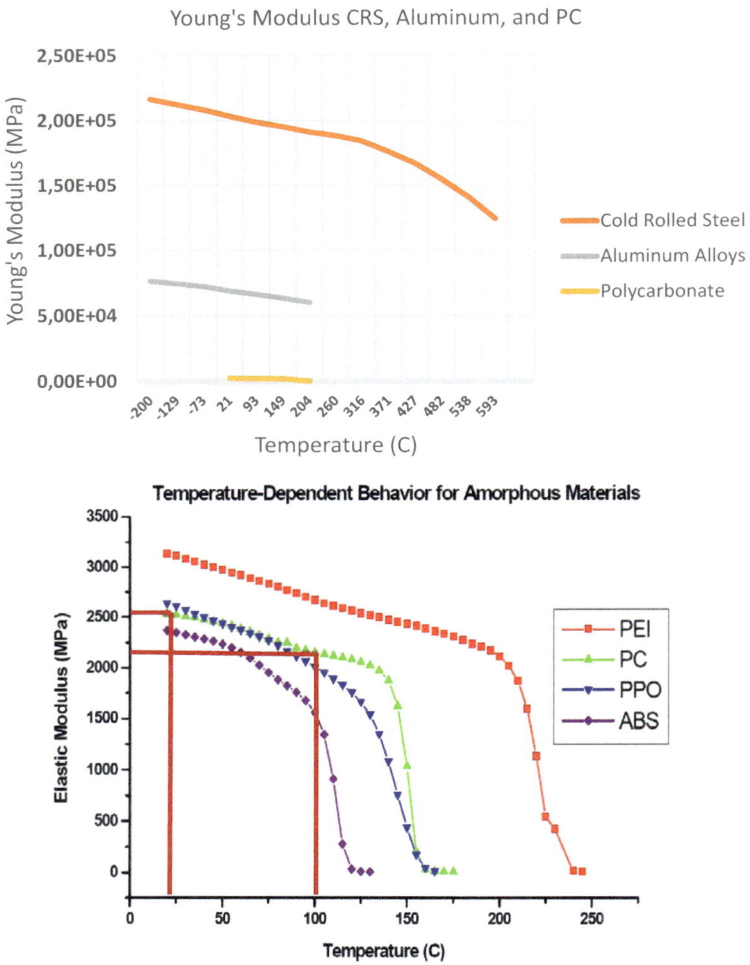

Figure 2.41 Temperature-dependent behavior of some common amorphous materials vs. common alloys [9]

2.3.5 Latent Defects

Metal parts contain very few defects that cannot be detected by conventional quality control equipment. An example is cracks at bends in stampings. The most common latent defects are porosity in castings and diecastings. Even the hidden porosity can be detected by X-Ray equipment. See Figure 2.42.

Figure 2.42 Defects in diecastings and stampings [8]

On the other hand, plastics may be full of latent defects that may not show up in any nondestructive testing. These defects will seriously impact the performance of the products both in the short and long terms. More in-depth discussion will follow in the following chapters. For now, suffice it to say that the designer has to keep this in mind when considering the performance of plastics.

2.4 Choosing the Right Plastic for Your Application

There are currently over 85,000 individual plastic formulations and counting. How does one choose the right needle in this hay stack?

Let us start with a look at the overall thermoplastics family.

2.4.1 The Commercial Plastics Family

Figure 2.43 provides a very high level view of most of the commercial plastics available today. The predominant performance distinguisher is the heat resistance of the plastics, generally described as the HDT (heat deflection temperature) and the continuous use temperature. The other Y axis shows the relative volume in use.

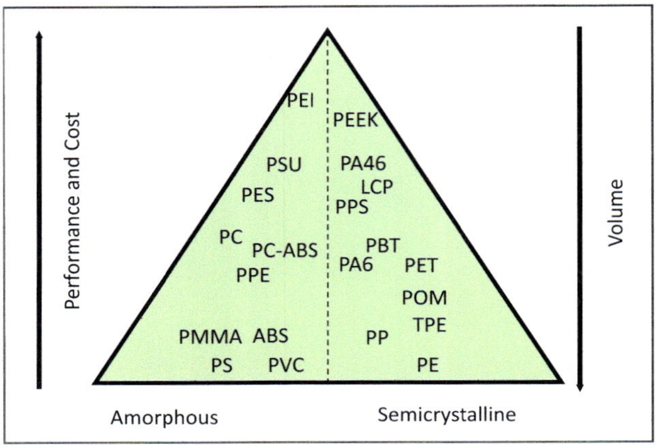

Figure 2.43 High level classification of plastic resins

Table 2.3 provides the properties of some high temperature resins.

Table 2.3 Thermal Properties of High Temperature Plastics

Polymer		Morphology	Glass Transition Temperature [°C]	Continuous Use Temperature [°C]	Heat Deflection Temperature [°C] with 40% Glass at 1.82 MPa
Polyphenylene Sulfide	PPS	Semicrystalline	92	204-232	260
Polyetheretherketone	PEEK	Semicrystalline	143	232-260	316
Polyetherketone	PEK	Semicrystalline	165	260-288	>316
Polyphthalmide	PPA	Semicrystalline	125	177-204	279
Polyetherketoneketone	PEKK	Semicrystalline	166	232-260	>316
Polyimide	TPI	Semicrystalline	250	260-288	332
High Temperature Nylon	HTN	Semicrystalline	125	149-171	260
Polysulfone	PSU	Amorphous	190	149-171	185
Polyethersulfone	PES	Amorphous	224	177-204	216
Polyetherimide	PEI	Amorphous	213	177-204	210
Liquid Crystal Polymer	LCP	Liquid Crystal	-	204-232	321

2.4.2 Material Selection Process

More than 60 polymer families plus hundreds of additives in existence today translate into over 85,000 grades and counting as stated earlier. How does one choose the one resin that will meet the performance and cost needs? [13]

The high level criteria are:

- Cost
- Resin morphology: amorphous or crystalline
- Mechanical properties

    - Modulus
    - Tensile and yield strength
    - Elongation
    - Toughness
- Thermal properties
    - Heat deflection temperature
    - Glass transition temperature T_G
    - Melt temperature T_M
    - High and low temperature requirements
- Chemical resistance
- Chemical compatibility
    - Overmolding
    - Proximity to other plastics
- Long term properties
    - Creep
    - Stress relaxation
- Special characteristics
    - Low friction
    - Fatigue
- Optical properties
    - Transparency
    - Refractive index
    - Birefringence
- Regulatory requirements
    - UL flammability
    - RoHS
    - Country and industry specific requirements such as FDA
- Molding properties: flow length
- Electrical properties
    - Resistance
    - Dielectric properties
- Environmental properties
    - Outdoor exposure
    - Corrosive environment

2.4.2.1 Morphology

Morphology refers to how the polymer chains are arranged in a certain polymer. As explained before, the two common arrangements are amorphous and semicrystalline. Table 2.4 highlights the high level differences between the two.

Table 2.4 High Level Differences between Amorphous and Semicrystalline Plastics

	Semicrystalline	Amorphous
Lower shrinkage		✓
Lower warpage		✓
Holding tight tolerances		✓
Filling thin walls	✓	
Chemical resistance	✓	
Transparency and optical requirements		✓
Fatigue/creep resistance	✓	
Abrasion/wear resistance	✓	

This is the starting point in the material selection process.

The choice is not always clear-cut. For instance, we may be designing a closed tolerance enclosure for a device that is used in an automotive shop. The obvious choice of using a tough amorphous material such as polycarbonate may not work because of its poor chemical resistance to oils and greases. In this case, we may have to choose a semicrystalline material such as glass-filled nylon and find other ways to overcome its inherent warpage and uneven shrinkage issues.

On the other hand, if the part has to be optically clear, only a few materials such as PC, PMMA, or polystyrene may be available in the order of toughness. If one is designing a high durability living hinge, the choice may be only different grades of polypropylene. If high fatigue is a requirement, the natural choice would be acetal.

It goes without saying that cost is the second important consideration in the selection.

My old professor used to say, “A good design is one that performs as designed AND costs as little as possible.” In that vein, the designer has to keep the cost in mind. A TV enclosure that can withstand the normal use requirements in a high impact polystyrene (HIPS) should not be designed in PC, which may cost three times or more.

Take another example–the common clear cold drink cup made out of polystyrene. One could use polycarbonate for the application, which would make it almost impossible to break and suitable for hot coffee. The cost, however, would be significantly higher both in terms of material and processing.

Can one justify the higher cost?

Figure 2.44 shows the relative cost of the more common resins.

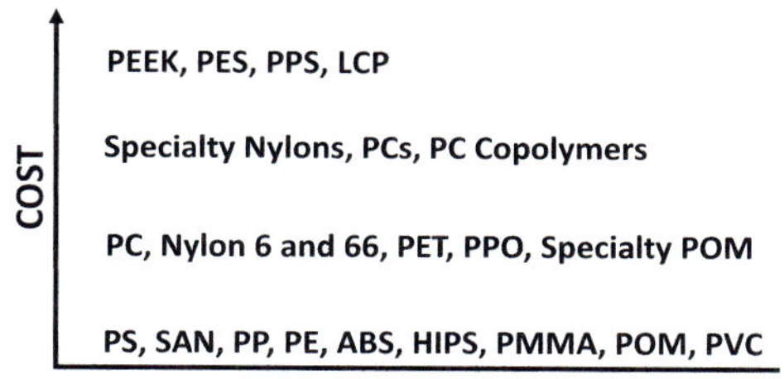

Figure 2.44 Comparative cost of resins

2.4.2.2 Thermal and Mechanical Properties

Next we look at the thermal and mechanical properties.

As discussed before, for most resins, cost and thermal properties are almost proportional. Thus in Figure 2.44, we could replace cost with thermal properties and come out with more or less the same order.

Table 2.5 lists the high level tensile strength of common generic resins. The values for nylon are for glass-filled ones. The properties for unfilled nylons would be around 12,000 to 15,000.

Table 2.5 Comparative Tensile Strength of Common Resins

Amorphous	Semicrystalline
Amorphous Nylon - 27000	Nylon 6-6 - 26000
PC - 9000	Nylon 6 - 23000
SAN - 10500	Nylon 12 - 23000
PMMA - 9400	PBT - 8500
ABS - 7000	POM - 8700
PS - 7700	PP - 5000
	HDPE - 3500

2.4.2.3 Special Properties

Plastics are not metals. And while plastics cannot compete with metals on tensile strength, modulus, longevity, electrical and thermal conductivity, etc., they have unique properties of their own that metals do not. Some of these are listed below.

2.4.2.3.1 Toughness

Toughness is one of the strongest advantages of plastics over metals. That is the reason why metal bumpers have been replaced by plastic ones in automobiles and (especially) industrial handheld devices like rugged mobile phones today have plastic enclosures. Although a lot of the smartphones now have partial or full metal enclosures for the sake of appearance, they cannot compare in performance and cost with the ruggedness of plastic enclosure. The terminal mentioned in Chapter 1 would not have been nearly as rugged if made out of machined or die cast metal.

Figure 2.45 shows a very popular smartphone with the back made out of machined aluminum. Notice the very visible dent from everyday handling that would not have been an issue in a well-designed plastic enclosure.

Figure 2.45 Dented cover of popular smartphone made out of machined aluminum

That being the case, a special discussion on this property is warranted. Even more so, since the measure of toughness in supplier property tables can be misleading.

Toughness is the measure of the energy a sample can absorb before it breaks. Let us look at Figure 2.46 (a), which represents the stress and elongation at the point of break. The cross-hatched area between the strain and stress curve is the toughness, or the energy required to break the sample. From simple physics, the stress is the force and the strain is the elongation at break (distance). Force × Distance = Energy.

Now let us look at Figure 2.46 (b). Starting from the left, the first curve represents a strong (high tensile strength) plastic that is not tough (brittle), such as polystyrene or PMMA. The middle curve represents a plastic that is strong and tough, such as polycarbonate; and the last curve represents a plastic that is neither strong nor tough, such as low density polyethylene or wax.

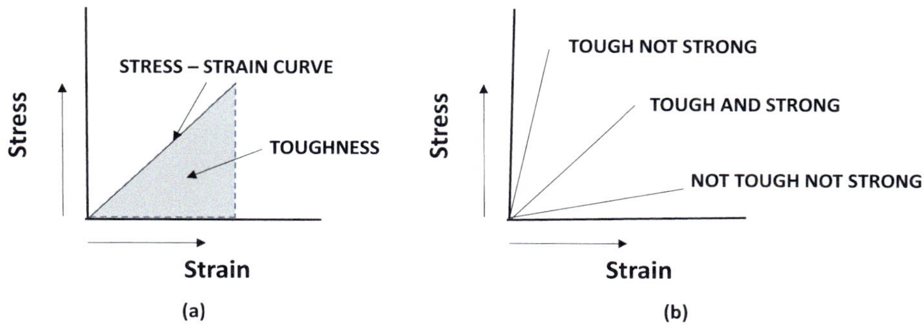

Figure 2.46 The shaded area in (a) shows the energy at break [14]; (b) shows curves indicating relative toughness and strength

In general, semicrystalline materials are tougher than amorphous materials. However, there are exceptions like PC, which is one of the toughest commercial and reasonably priced materials available to designers.

Other properties important in plastics are:

2.4.2.3.2 Coefficient of Friction

Here again almost without exception, hard semicrystalline materials such as nylon and acetals have much lower coefficients of friction than their amorphous counterparts, such as PC and acrylic. PTFE (Teflon™) is a ubiquitous example of a low friction material.

2.4.2.3.3 Optical

No metal can transmit light. Among plastics only amorphous plastics can transmit enough light to be termed transparent. The best that semicrystalline materials can be is translucent. Nevertheless, in some semicrystalline materials like polypropylene, nucleating agents help to improve the transparency.

There are two main criteria for optical properties:

1. Light transmission (in percentage)

   Obviously, the higher the number, the better the transmission. Acrylics, polystyrene, and polycarbonates have light transmission from the high 80s to the low 90s (percent). This is comparable to clear glass.

2. Refractive index

   Everyday requirements for the human eye to see through picture frames, glazing, clear cups, etc. can be met with common resins such as PS, PC, and acrylic.

However, in the process of molding the degree of induced molded-in stresses can make the rays going through the part bend in uncontrolled directions. This is termed birefringence. This fact makes these plastics impractical for optical systems that require the paths of the rays to go through minimum bending, such as light guides and devices for bar code scanning.

Special optical materials that can provide satisfactory optical properties are available, such as TOPAS COC (Cyclic Olefin Copolymer) from Polyplastics with minimal birefringence. Their mechanical properties, however, are very low.

2.5 Thermoplastic Elastomers and Overmolding

The plastics we have been dealing with so far are rigid plastics, used in applications where stiffness is an essential performance requirement. There is another class of thermoplastics called thermoplastic elastomers (TPE). These are low modulus, flexible materials that can be stretched repeatedly at room temperature and return to their approximate original size when the stress is released. These are used for applications requiring:

- Greater flexibility and compressibility and "rubberiness," such as seals, bumpers, soles for shoes, cable insulation, etc.
- Soft touch feel and aesthetic appeal for rigid plastics being used by human hands. These can be
  - Standalone components
  - Components overmolded over rigid plastics or even metals–as in handheld electronic devices
  - Components mechanically fitted over rigid plastics or metals, such as plier handles
- Sound dampening

As in the case of rigid thermoplastics, there is a large family of thermoplastic elastomers.

2.5.1 TPE Family

The most common families [10] are:

- SBC (Styrenic Block Copolymers)
  - SEBS (Styrene Ethylene/Butadiene Styrene): primary product in the SBC category
  - SEPS (Styrene Ethylene/Propylene Styrene)
  - SBS (Styrene Butadiene Styrene): inexpensive (pencil erasers, footwear, etc.)

- TPO (Thermoplastic Olefins)
- TPV (Thermoplastic Vulcanizate)
- TPU (Thermoplastic Polyurethane)
- MPR (Melt Processable Rubber)
- COPE (Co-polyester Ether; TPC-ET)
- PEBA (Polyether Block Amide)
- COPA (Co-polyamide Elastomers)

All of the above contain a hard and a soft block. As an example, SBS has styrene as the hard block and butadiene as the soft block.

The hard block contributes to plastic properties such as:

- Processing temperatures
- Continuous use temperature
- Tensile strength
- Tear strength
- Chemical and fluid resistance
- Adhesion to overmolding substrates, inks, adhesives

The soft block results in elastomeric properties such as:

- Lower service temperature limits
- Hardness
- Flexibility
- Compression set and tensile set

Cost

Figure 2.47 shows the comparative costs of TPEs.

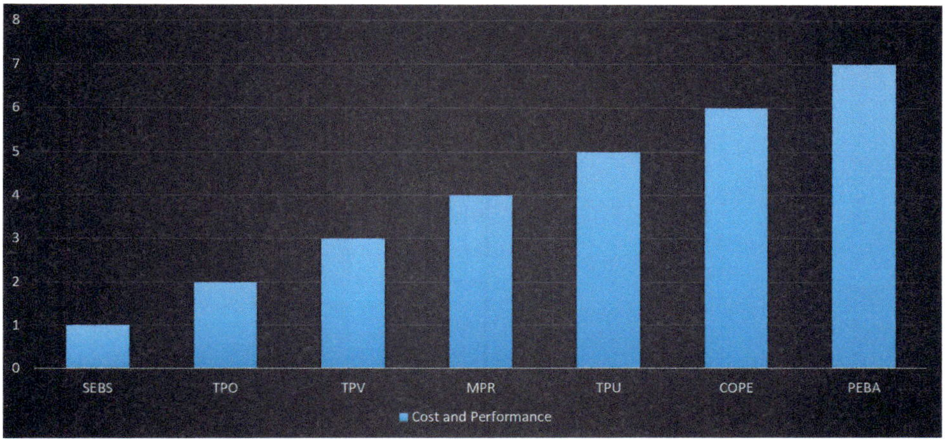

Figure 2.47 TPE cost and performance [10]

2.5.2 Compatible Materials

Not every TPE will be bonded with or can be molded with every hard resin. The selection needs to be based on:

1. Thermal compatibility

   The molding temperature of the TPE should be in the range where it will not make the substrate "melt," causing it to lose its basic shape and/or cause an unacceptable distortion upon ejection.

2. Bonding

   The overmold should adhere well to the substrate based on chemical compatibility and other adhesion factors. Discussion on this matter with the TPE suppliers is highly recommended.

As an example, the following materials from PolyOne, Inc. [15] indicate their compatibilities with common hard plastics.

Versaflex™ CL 2242 and CL 2250 – Clear bondable grades for PP overmolding

Dynaflex™ G2706 – Soft translucent grade for PP overmolding

Versalloy™ HC 9210 & 9220 Series – Natural, colorable grades for PP overmolding

Dynaflex™ G2711 – Highly colorable solution for PP overmolding

Versaflex™ OM 3060 – Silky clear solution designed for overmolding onto polycarbonate (PC), acrylonitrile butadiene styrene (ABS), and PC/ABS blends

Versaflex™ OM 1040X – Highly colorable solution for overmolding onto PC, ABS, and PC/ABS blends

All other TPE suppliers have similar information for their own resins.

2.6 References

[1] Images provided by courtesy of SABIC

[2] This section includes material adapted with permission from Joseph P. McFadden Sr.'s articles on the subject

[3] D.L. MacAdam; *Visual sensitivities to color differences in daylight* (abstract), JOSA 32 (5): 247–274, doi:10.1364/JOSA.32.000247 (May 1942)

[4] MacAdam ellipses for one of MacAdam's test participants, Perley G. Nutting Jr. (observer "PGN"), plotted on the CIE 1931 xy chromaticity diagram. The ellipses are ten times their actual size, as depicted in MacAdam's paper.

[5] Adapted from *Basics of Color Perception and Measurement* (c), Hunter Lab (2001)

[6] Adapted from *http://www.dc.engr.scu.edu/*

[7] Photo credit: Joseph P. McFadden Sr.; *http://McFaddenCAE.com*

[8] X-Ray Inspection of Castings (Army Air Force's Requirements), Robert Katz, University of Nebraska (non-copyrighted), University of Nebraska-Lincoln, rkatz2@unlnotes.unl.edu

[9] Courtesy of Michael Sepe

[10] Courtesy of Mike Walter, Bay State Polymer. Extracted and modified from Paper at ANTEC 2015.

[11] Joseph P. McFadden Sr.; *http://McFaddenCAE.com*

[12] *https://annoyzview.wordpress.com/2011/12/27/rust-free-iron-pillar-of-qutub-minar/*

[13] Mukund Patankar has contributed extensively to this section

[14] Adapted from *http://pslc.ws/macrog/mech.htm*

[15] *http://www.glstpes.com/products/thermoplastic-elastomers/kraton-thermoplastic-elastomers*

3 Design

3.1 Introduction

The designer of a product is like the quarterback in American football. The quarterback is responsible for the outcome of each individual play and the ultimate outcome of the game. Even the strongest offense and defense team members cannot make up for the deficiency in the initial play by the quarterback. Similarly, the designer controls the ultimate performance of her/his product. The rest of the members of the team–resin supplier, tool-maker, and molder–cannot make up for the deficiencies in the design.

Going back to David Wright's causes for plastic parts failure (Figure 1.1, Chapter 1), his study found that 20 % of the failures of plastic parts could be assigned to the part design. As I will demonstrate in the pages to come, from my experience, a vast majority of the **apparent** material, tooling, processing, and even abuse issues can ultimately be traced back to poor design.

Let me also preface the chapter with this statement:

The discussion assumes that the designer is familiar with the normal stress and strain calculations. There is no need to add pages to the book on this area as they are readily available in hundreds of excellent books on mechanical design and the internet.

Ultimately, calculations for stress, strain, wear, etc. on a component are the same regardless of the material being used. As an example, the stress on a certain member is P/A (where P is the applied load and A is the cross-sectional area), regardless of the material.

The following pages will also not go into specific features such as gears, screw threads, snaps, etc. Again, there is already a lot of excellent existing literature in this area. I am, however, going to provide useful references for all of the above areas, which I have personally used as a guide to good design. Good news–many of these sources are free!

3.2 Robust Design

Before we start, let us agree on the definition of robust design. In my view, the following are the requirements for a robust design:

A product with a robust design should:

- Meet or exceed the requirements over the projected life and intended environmental conditions
- Meet the appearance requirements
- Be as cost effective as possible

Let us start with the simplest requirement–cost.

3.2.1 Cost

Let us assume that the total number of products to be produced is n. Therefore, the total cost of producing the part over its lifetime is the sum of the legitimate costs, the avoidable costs, and the beyond avoidable costs.

3.2.1.1 Legitimate Costs

These consist of:

3.2.1.1.1 Normal Development Costs

These are the costs to design the product, which include items such as:

- CAD time (initial and based on simulation and testing)
- Flow simulation[1] and analysis time (FEA, Flow)
- Prototyping
  - Appearance (hand-crafted, cast polyurethane, etc.)
  - Spatial (machined, stereolithography, etc.)
  - Functional (soft-tooled–note that I am NOT including machined parts here. Please see explanation in Chapter 2, Section 2.3.3.1)
- Testing
- Documentation: solid models, 3D and 2D drawings, bill of materials (BOM), etc.
- *Reasonable* engineering changes: I define reasonable engineering changes as the ones resulting from product requirements changes beyond the control of the designer, changes indicated by physical testing that were beyond the scope of reasonable and diligent analysis and simulation, etc.; these do not include

[1] Notice that flow simulation is at the design stage. This is very important. More in Chapter 9, Section 9.6.

changes due to negligence or lack of awareness of design fundamentals in the initial design

- Tooling and manufacturing reviews

3.2.1.1.2 Tool Cost

The tool cost consists of:

- Cost of designing and building the tool(s)
- Tool trials and initial corrections to the tools
- *Reasonable* engineering changes
- Tool maintenance costs–routine and due to wear

3.2.1.1.3 Part Cost

Total cost of parts:

- Material cost: n×(weight of part, runners, reasonable scrap)×price per unit weight
- Processing cost (machine and operator(s)): n×(cycle time in seconds)×labor and machine cost per second
- Reasonable scrap costs
- Reasonable profit
- Normal quality assurance, handling, and storage costs

3.2.1.2 Avoidable Costs

Additional resources spent to fix errors + additional tooling costs + scrap cost

3.2.1.3 Beyond Avoidable Costs

3.2.1.3.1 Cost of Fixing Errors

Figure 3.1 is the generally accepted curve for the cost of fixing errors in various stages of design and it hardly needs an explanation.

Just as a reminder, let us talk briefly about the recent issue with the ignition switch on the General Motors cars and how very simple design errors can lead to huge losses. Here are some quick statistics:

The cost for getting the switch designed right in the first place would have been 57 cents [1]. The total cost of the recall was $4.1 billion as of February 4, 2015 [2].

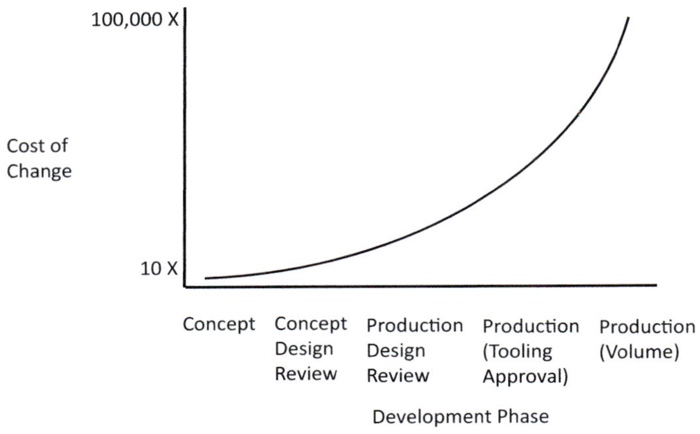

Figure 3.1 Cost of fixing errors

3.2.1.3.2 Opportunity Costs

Figure 3.2 is another commonly accepted curve for the typical cost of the life cycle of a typical product.

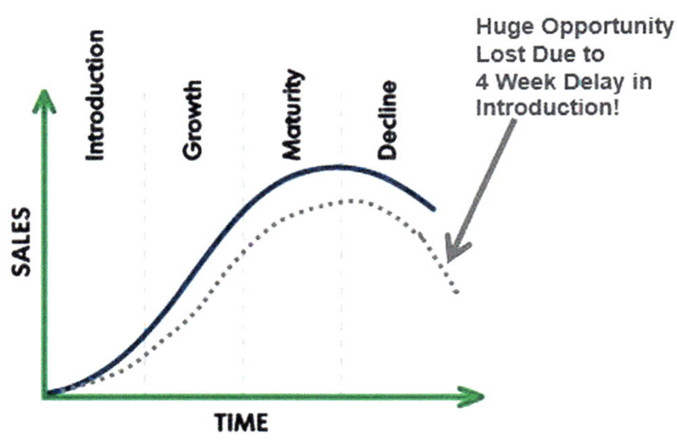

Figure 3.2 Opportunity cost associated with delayed release

Some products, such as a smartphone, can have a very short life span. Others, such as a commercial jetliner, can have a very long life span. The Boeing 737 was introduced in 1967. Forty-five years later, in July 2012, the 737 became the first-ever commercial jet airplane to surpass the 10,000 orders [3].

The integral of the solid line in Figure 3.2 shows the total anticipated sales for a product over its intended life. The dotted line represents what may happen if the product is released just four weeks later than the planned introduction. The difference between the integrals of the two curves potentially represents the total loss or the total opportunity costs.

Using the example of commercial jetliners again, the grounding of the Boeing 787, because of the battery overheating, cost Boeing an estimated $600 million in direct costs, halted deliveries, and forced some airlines to lease alternative aircraft [4]. As large as this figure is, it may pale in comparison to the lost opportunity costs. At about $200 million per aircraft, the sales of just 5-6 planes lost to Airbus may mean a revenue loss of one billion dollars–forever!

Let me give you a couple of my own examples of how simple plastic failures can cause losses of hundreds of thousands of dollars or more. In the early eighties I worked at Exxon Office Systems in Pennsylvania, which produced cutting-edge word processors. The high-end devices cost upwards of $10,000 per unit (in 2016 dollars, about $35,000!). On one of the models, I was the engineer in charge of ultrasonically welding a clear plastic part called format scale (costing all about 50 cents) to the typewriter housing. Unknown to me, the molder changed the molding process, which resulted in the format scale not being able to be welded to the housing. I was ready to leave for a long weekend on a Friday afternoon when I received a call from the plant manager asking me in a not too polite tone to get to his office in a hurry. Instead of having the wonderful getaway that I was planning on, I was shuttling between our plant and the molder trying to remedy the situation for the next few weeks. In the meanwhile, the assembly line came to a complete halt!

If that was not bad enough, I was also responsible for about a four-week delay in the introduction of a newer, fancier model because the ABS top cover on the keyboard I had designed was warping in the high temperature storage test. One plastic part that cost under a dollar prevented the shipment of word processors with a price tag of over ten thousand dollars for over four weeks. (More on the above two examples in the pages to follow.)

Beware! As a plastic product designer, you may be the one that causes the four-week delay in Figure 3.2 due to a simple failure, such as a *preventable* broken snap in a plastic part unearthed in the final rounds of testing!

3.2.2 Performance, Manufacturability, and Costs

The following are the most common requirements for designing a plastic part:

- Purely physical requirements:
  - Mechanical: tensile strength, toughness, elongation, thermal, etc.
  - Electrical: dielectric strength, resistance, etc.
- Chemical and environmental requirements

- Agency and regulatory requirements
- Molding and tooling dependent requirements:
  - Wall thickness uniformity
  - Wall thickness ratio
  - Flow direction–from thick to thin
  - Maximum and minimum wall thickness
  - Eliminating sharp internal corners
  - Adequate draft
  - Boss lengths–unsupported cores and rib heights
  - Knit line–location, strength, and appearance
  - Effect of orientation
  - Gate size and location
  - Long term stresses
  - Thermal compatibility
  - Material compatibility

3.2.3 Purely Physical Requirements

While a lot of physical property requirements and stress and strain calculations may be similar to the ones for metals, there are many that are not. As I stated before, I am not going to discuss the conventional stress and strain calculations. I will, however, dwell on the key differences. Also stated before, information is provided in the reference section (Chapter 9) for designing features such as snaps, gears, etc.

3.2.4 Chemical and Environmental Requirements

In the metals world, this was fairly predictable and a lot of data is available. As stated in Chapter 2, in addition to the inherent chemical and environmental resistance of the material itself, design, tooling, and processing can enhance this capability with the elimination of as many molded-in or applied long term stresses as possible. The reverse will weaken the part in this capability.

3.2.5 Agency and Regulatory Requirements

Some are common with metals; some are unique to plastics. These may be related to:

- Safety, including mechanical, electrical, and fire prevention, such as UL
- Medical and food contact requirements, such as FDA
- Environmental requirements, such as RoHS

The designer needs to be aware of the above requirements in material selection, design, tooling, and processing and make sure they are met.

3.2.6 Tooling and Molding Related Requirements

These go beyond just conventional DFM (design for manufacturing) requirements. These are also fundamental to the eventual performance of the product.

3.2.6.1 Wall Thickness

The main wall of any plastic part is called the nominal wall. The nominal wall has three basic functions:

- Perform the functional and visual requirements as a structural member of the product
- Support all the individual features that are attached to or a part of it
- Act as an efficient conduit to carry the plastic from the gate to the extremities of the part

3.2.6.1.1 Maximum Wall Thickness

This is one of the biggest differences between metal and plastic parts. In the metals world, with few exceptions, the stress carrying capacity is proportional to the wall thickness.

As an example, in the case of a directly applied load, if P is the maximum load carrying capacity and t is the thickness of the wall, in **all cases**,

$$P \propto t \quad (3.1)$$

Therefore, the greater the thickness, the greater the stress carrying capacity of the component.

In the case of plastics, however, thicker is not necessarily stronger. Referring to the discussion in Chapter 2, Section 2.3.3, the molding phenomenon creates compressive stresses in the outer layer and tensile ones in the middle. Since most induced stresses are in tension, we need to maximize the layers in compressive stress and minimize the ones in tension. A thick part will have relatively thin layers of compressive stress compared to the layers under tensile stress.

For *conventional* molded parts, there is a maximum thickness beyond which there will be a loss in the mechanical properties.

Let us talk briefly about conventional, foamed, and thin walled parts. Most moldings can be divided into three wall thicknesses:

- Thin wall–generally 1 mm and less
- Conventional–between 1 to 3.5 mm
- Thick–over 3.5 mm thick–with air or foam introduced in between the solid outside layers; the most common ones being gas assisted or structural foam

Most of the discussion to follow will be on conventional wall molded parts. Gas assisted and structural foam moldings are specialized areas that I will provide reference material for in Chapter 9. Differences in the design rules for thin wall parts will be highlighted where appropriate.

For the most part, material, tooling, and processing considerations are similar in all three cases.

The thickest wall is based on the best compromise between structural, processing, and agency requirements.

3.2.6.1.1.1 Critical Thickness

Generally, increasing wall thickness reduces deflection during impact and increases the energy required to produce failure up to a certain limit. Many materials, such as polycarbonate, lose impact strength if the thickness exceeds this limit, known as the critical thickness. Above the critical thickness, parts made of polycarbonate can show a marked decrease in impact performance.

Walls with thickness greater than the critical thickness may undergo brittle, rather than ductile, failure during impact.

Referring to Figure 3.3, the impact strength falls rapidly at −20 °C from over 12 ft·lb/in to about 3 ft·lb/in. The part basically becomes very brittle. Considering that −20 °C is not that uncommon in many different areas of the world and also taking into account a reasonable factor of safety, the maximum thickness for this material should be less than 3.5 mm (0.140 in).

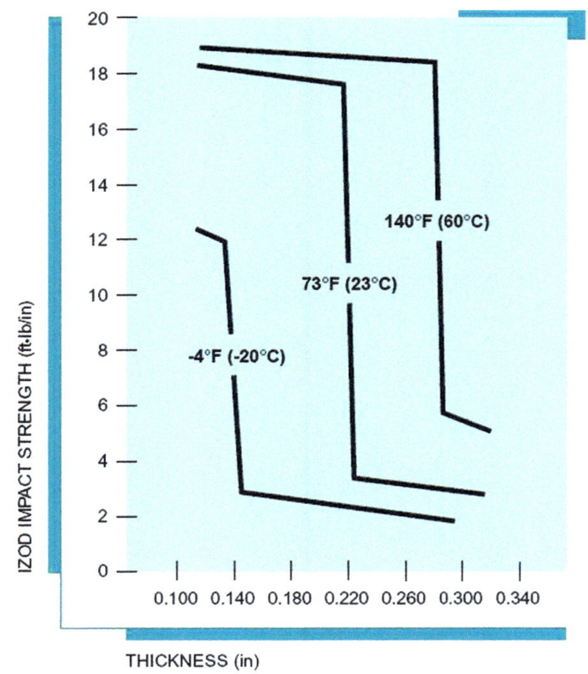

Figure 3.3 Drop in impact strength with increase in thickness [5]

3.2.6.1.1.2 Processing Cost

The cooling portion of the molding cycle time is roughly proportional to the square of the thickness.

Looking at Figure 3.4, the cooling time for a wall thickness of 2.0 mm is about 3 seconds. For 4.0 mm it is about 12 seconds approximately (or 3×2^2). The molding rate for a small tonnage machine in the U.S. may cost \$30/hour. The extra approximate 9 seconds cost \$30×9/60/60 = \$0.075 or 7.5 cents. While this may be an increase of only 10% for a large part costing about 75 cents, it may double the price of a part costing 8 cents!

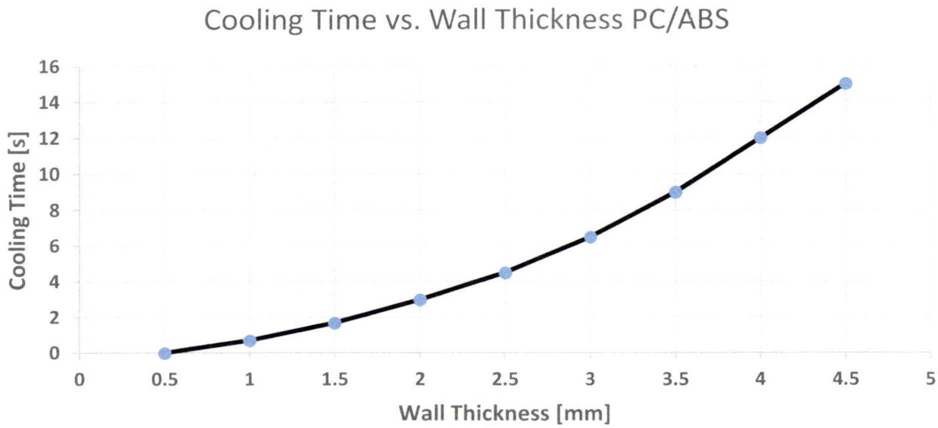

Figure 3.4 Cooling time vs. wall thickness

Here is an example:

Figure 3.5 shows the ubiquitous latch for the folding tray table found on many commercial planes. The section in the middle is not cored out as far as I can tell. Using ABS for the material, if this were molded on a single cavity mold, the cost of the *cooling portion* of the molding cycle would be about 2.5 cents (3 seconds cooling time) if the section were cored out. Because it is not cored out, the cooling time is approximately 12 seconds with a cost of 10 cents–about 7.5 cents more. If the material were ABS at about $2.50/kg and the part weighed 10 g, the material cost would be about 2.5 cents. With a 15 seconds *total* cycle (12.5 cents), the cost of the cored part would be 2.5 + 12.5 = 15 cents. With the additional cooling time of 9 seconds, the cost goes up by 7.5 cents for a total of 22.5 cents or 50%!

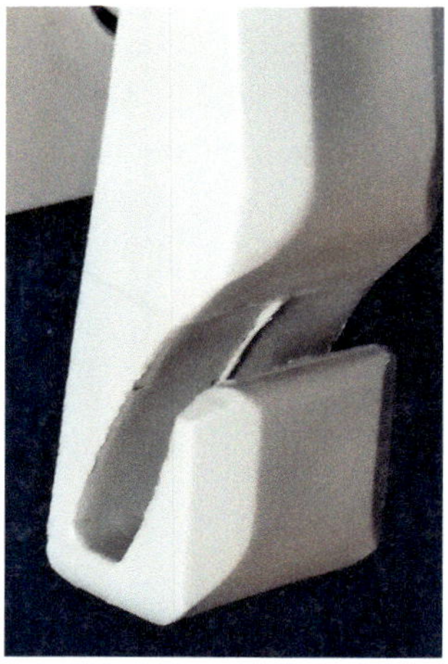

Figure 3.5 Airplane folding tray latch

3.2.6.1.2 Minimum Wall Thickness

The minimum thickness of the wall depends on four factors and should be the lowest of these:

- Thickness required for flow to the extremities of the part
- Thickness required to have the proper ratio with the thickness of ribs in the part
- UL flammability requirements
- Structural requirements

Let us examine these in detail.

3.2.6.1.2.1 Thickness Required for Flow to the Extremities of the Part

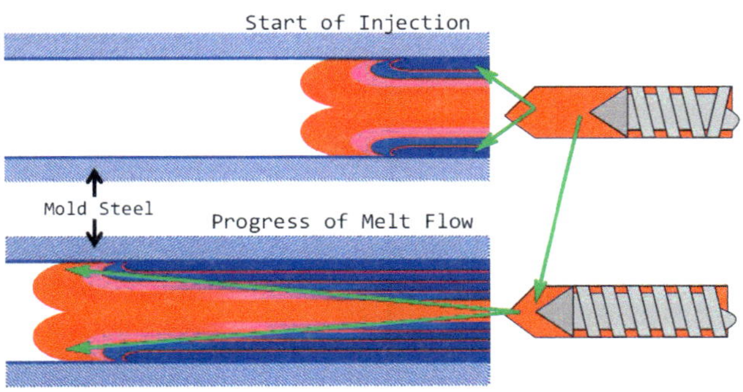

Figure 3.6 Fountain flow [10]

The common term for the flow of plastic from the gate to the cavity is fountain flow. Looking at Figure 3.6, the material entering first "freezes" against the mold wall and the next internal layers flow to fill up the succeeding wall area. The frozen area next to the wall continues to become thicker at the start of the flow, thus making the channel available for flow continuously smaller. Additionally, during the flow of plastic from the gate to the extremities of the part, the flow front also continues to cool down.

As the temperature goes down, the material becomes more viscous. At some point the combination of the available pressure and the remaining flow cross-sectional area is not large enough to make the plastic flow any further. It stands to reason, therefore, that the larger the wall thickness, the longer the flow path. Resin suppliers generally provide this data as shown in Figure 3.7.

Looking at Figure 3.7, from Covestro, material 2800 will flow for a total of a little over 10 inches (~260 mm) at a section thickness of 2 mm and about 22 inches (~560 mm) at a thickness of 3 mm. Therefore, if the flow length from the gate to the extremity of the part is 10 inches, the minimum wall needs to be at least 2 mm.

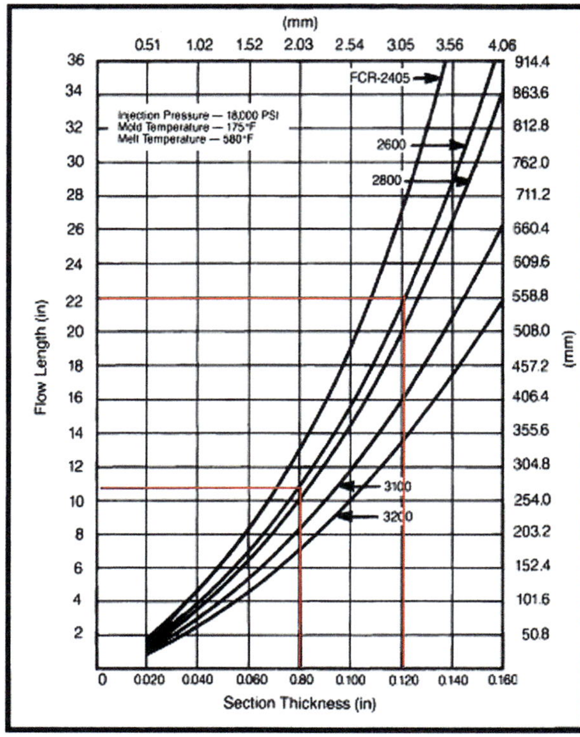

Figure 3.7 Thickness vs. flow length [5]

Flow simulation tools such as Moldflow, Moldex, etc. can also accurately predict the flow length of the part, as shown in Figure 3.8.

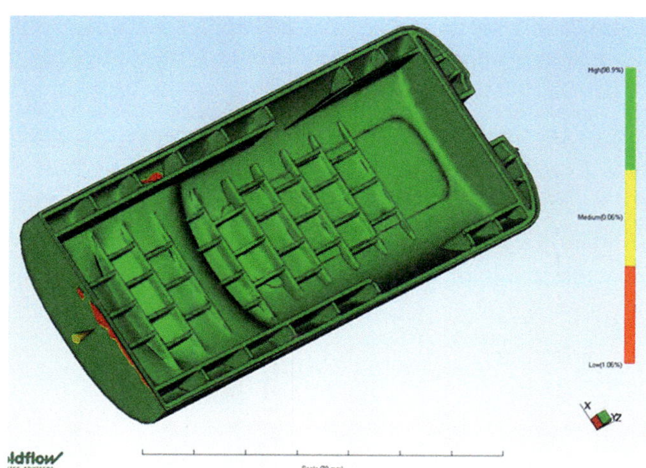

Figure 3.8 Flow length prediction by Moldflow™

Figure 3.8 shows the confidence of flow (green through red). The gate location is shown by the yellow cone. In this case the designer would need to increase the nominal wall thickness to make sure the ends of the part will be filled.

3.2.6.1.2.2 Thickness Required to Have the Proper Ratio with the Thickness of Ribs

See Figure 3.9. The thick horizontal wall is the nominal wall with a thickness *t*. The vertical wall is a rib protruding from the nominal wall. This becomes the nominal wall for the horizontal rib protruding from it. As will be discussed later in this chapter, the rib thickness should be half the thickness of the nominal wall for a proper design. Therefore, the vertical rib should be 0.5×*t* and the horizontal rib should be 0.25×*t*. If the designer started with a nominal wall of 2 mm, the thickness of the horizontal rib would be 0.5 mm. In most cases the material would not flow into this rib and a short shot would result. Therefore, for a design like this, the nominal wall thickness would need to be increased to make sure the horizontal rib fills.

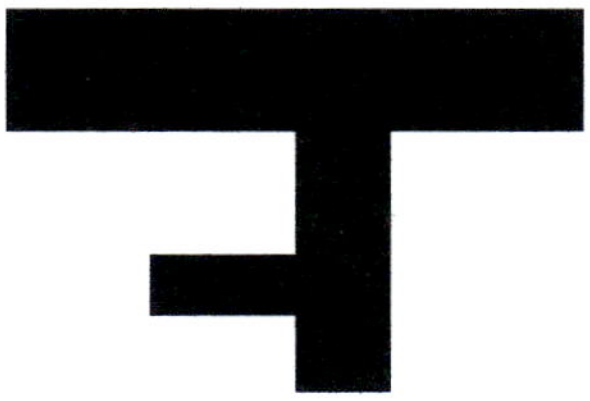

Figure 3.9 Rib within a rib (required inside fillets and draft removed for clarity)

Figure 3.10 shows a Moldflow simulation that predicts a similar situation when the required draft is included on the rib. Incidentally, it is very important that the draft be included from the very initial design to make sure the part will fill. *If a designer conducts spatial and structural analysis without a draft and finds out later that the part will not fill once the draft is included, it may be very difficult and costly to adjust the design at that point!*

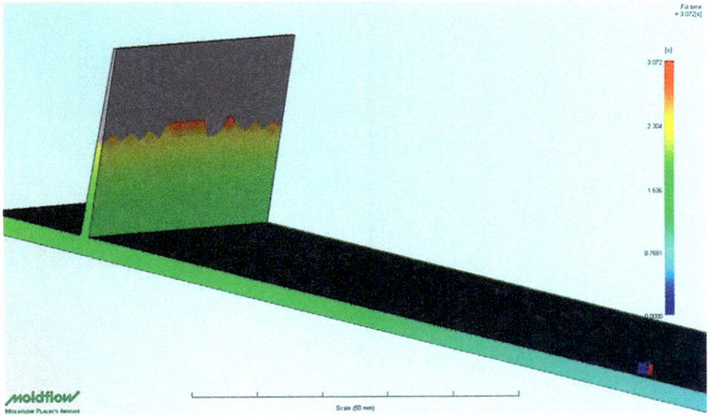

Figure 3.10 Moldflow simulation of no flow condition on long thin rib

3.2.6.1.2.3 UL Flammability Requirements

It is customary for manufacturers of electrical or electronic products such as computers, laptops, handheld communication devices, etc. to get a UL (Underwriters Laboratory) certification for the fire safety of their products. More information is provided in the appendix (Chapter 9) on UL and its certification. For now, let us take the example of UL 60950-1, Information Technology Equipment Safety. Let us assume that the walls of a safety enclosure need to have a UL94 rating of V0 (rate at which plastic burns). This rating requires a minimum nominal wall thickness.

As an example, the material selected is PC - SABIC Lexan 940. The SABIC literature indicates that the minimum wall thickness of a UL 94 V0 rating is 0.043 in. Therefore, the nominal wall has to be a minimum of 0.043 in. Caution—there may be additional requirements on other components of the enclosure. I would recommend that the designer speak with a UL subject matter expert before these components are designed.

3.2.6.1.2.4 Wall Thickness to Meet Structural Requirements

Finally, the minimum wall thickness needs to be one that meets the structural requirements based on the stresses various areas of the plastic part will see.

As an example, if the wall thickness of a snap at the base needs to be 1 mm in order to have a certain snapping force based on engineering calculations, and if the snap is protruding from the nominal wall, following the thickness of a rib from above, the wall may need to be only about 2 mm thick.

However, the wall may need to be increased 2.5 mm to meet its overall stiffness requirements.

3.2.6.1.3 Inside Corner Radius

In my experience, one of the leading causes of plastic parts failure is the lack of an adequate radius in inside corners.

Figure 3.11 shows the universal (plastics, metals, wood) relationship between the inside corner radius ratio (R/WT—WT being the wall thickness) and the *stress concentration factor*. A simple explanation of the concept of stress concentration is the relative intensification of an applied stress in a particular location of the part. A stress concentration of 1 would therefore indicate that the chances of a part failing at that location are no more than any adjacent area of the part. By the same reasoning, a stress concentration of 3 would indicate that the implied stress is concentrated 3 times more in that location compared to the adjoining area. Therefore, a part is three times likely to fail in that area than if the stress concentration did not occur.

Inside Radius and Stress Concentration

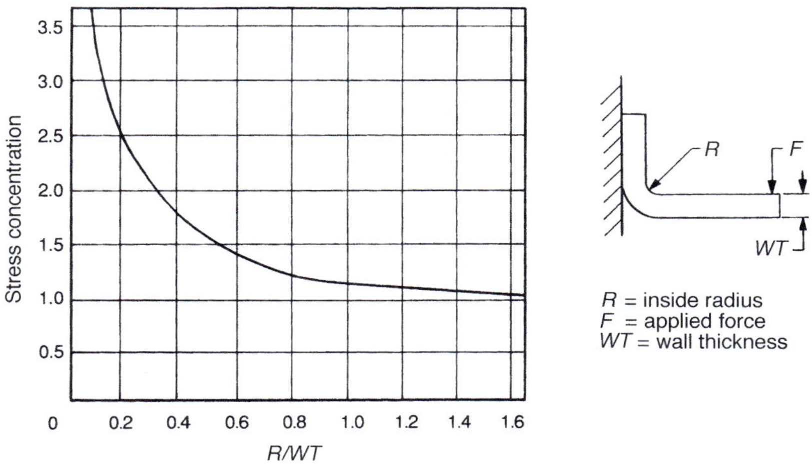

Figure 3.11 Relationship of inside corner radius to the stress concentration factor [6]

In maintaining a uniform wall thickness, there are practical limitations on how big this ratio can be without causing other issues. A good general balance is a minimum ratio of 0.5. It is important to remember that any radius is better than a sharp corner. As a side benefit, corners with internal radii improve the flow as compared to turning a sharp corner, resulting in a more uniform melt and possibly a shorter molding cycle.

Figure 3.12 shows the catastrophic failure in an otherwise very tough PC part in a short drop. Even before the drop, the sharp corners showed the white stress marks shown in the circle after the part was fastened to the mating part and disassembled.

Figure 3.12 Catastrophic failure of boss due to sharp internal corners in drop

As mentioned in Chapter 2, any stress in a plastic part will reduce the chemical resistance and increase the chance of ESC (environmental stress cracking) of the plastic part. It does not necessarily take a lot of external stress for a sharp internal corner to fail. Figure 3.13 shows a failure in the sharp corner of a microwave mounted over a stove. Although free of any applied stresses, the grease vapors from the cooking below caused a catastrophic failure in the area.

Figure 3.13 A microwave cover with cracked internal sharp corner

Here is another example to illustrate the point. Figure 3.14 shows severe cracks in the side of a blow-dryer. The cause was sharp internal corners on the inside of the cracked areas. The cracks were not caused by any harsh chemicals. Just the innocuous hand lotion and the moisture on the users' hands were enough to cause these severe cracks!

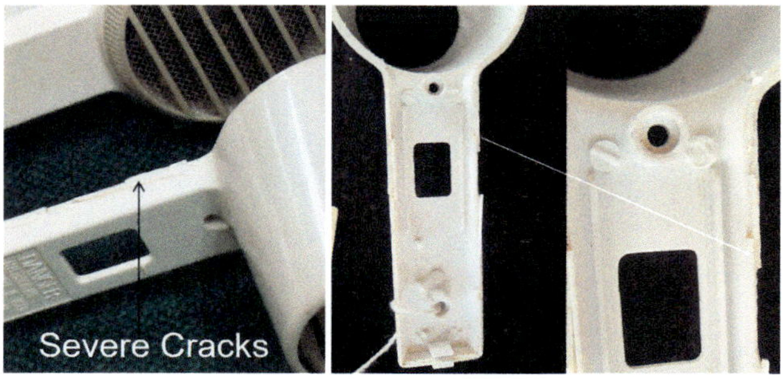

Figure 3.14 Severe cracks in the side of a blow-dryer

Figure 3.15 shows yet another example of sharp corners causing ESC failures without external forces applied. This is a handheld device made out of polycarbonate. Just the lotion and/or the sweat on the operators' hands caused the device to have severe cracks in the side opposite the sharp corners.

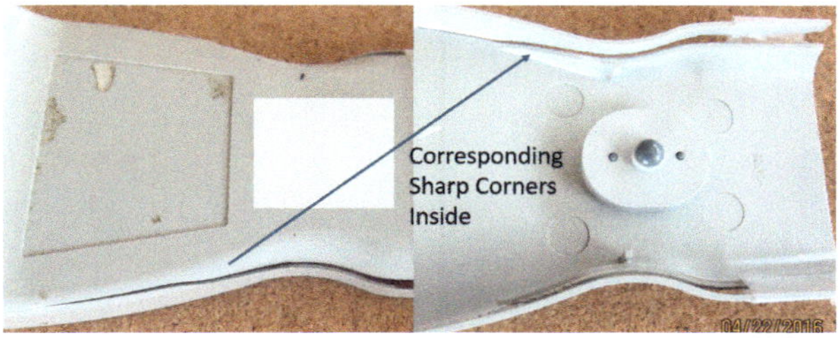

Figure 3.15 Severe cracks at sharp corners from attack by lotion and sweat on hand

3.2.6.1.4 Uniform Wall Thickness

This is one of the most important requirements of robust plastic parts.

Coming back to the theme of *plastics are not metals,* this is also one of the main characteristics that distinguish them from metals. In most metal parts, a uniform wall thickness is not a requirement. In fact, it is not uncommon to have metal parts consisting of heavy wall thicknesses next to a thin section. Diecasting, and castings in general, may be an exception but for a different reason. Unlike a thermal reason (to be explained shortly), it has to do more with flow; namely, feeding thick sections through thin ones is more difficult. As an example, imagine a 12 mm diameter hose connected to a 25 mm diameter hose. Water is continuously flowing from the smaller hose through the larger hose to an opening. The 25 mm hose will be only filled partially at any given time.

Now look at this requirement from the plastics point of view. Figure 3.16 shows the temperature at ejection of a part purposely designed by Glenn Beall to illustrate some common design issues and their effect on the performance of the part. He called it the DOs and DONTs block in his seminars going back to the 1980s.

This is an ABS part with the left half having half the wall thickness of the right. John Bozzelli, who in my mind is one of the biggest champions of scientific molding (more on this subject in the processing section, Chapter 5), took the infrared thermal image of the part as it was coming out of the mold. The left side has cooled down to 118 °C while the thick section on the right is at 238 °C. As time goes on, the left size will "freeze" and stop shrinking while the right will continue to do this for a very long time. At the interface it will introduce a very high stress, which will cause the part to warp in the short term and lead to structural weakness and poor chemical resistance in the long run. The part will also cost substantially more as a result of the longer cycle time (see earlier discussion on this subject) for the part to cool down to the safe ejection temperature. Even more additional cooling time may be needed to reduce the warpage.

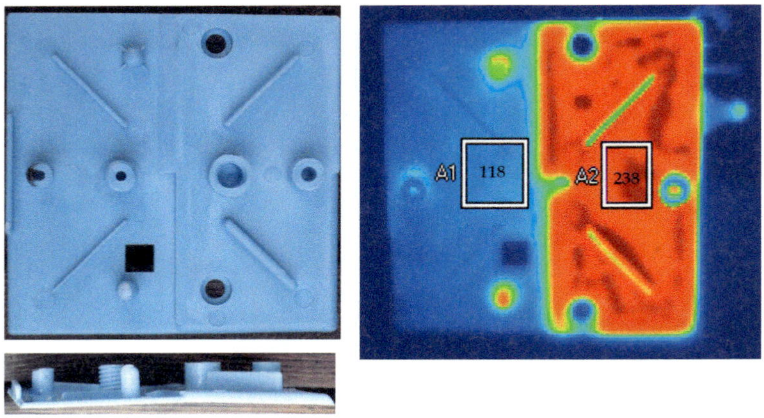

Figure 3.16 Surface temperature of plastic part at ejection [7]

Figure 3.17 shows the in-cavity residual stresses in a part with a non-uniform wall. The normalized wall thickness is 0.097 in or about 2.5 mm. The effective section thickness in the circled area is about 0.150 in or 3.75 mm due to the boss being attached to the wall next to it creating a thick cross section.

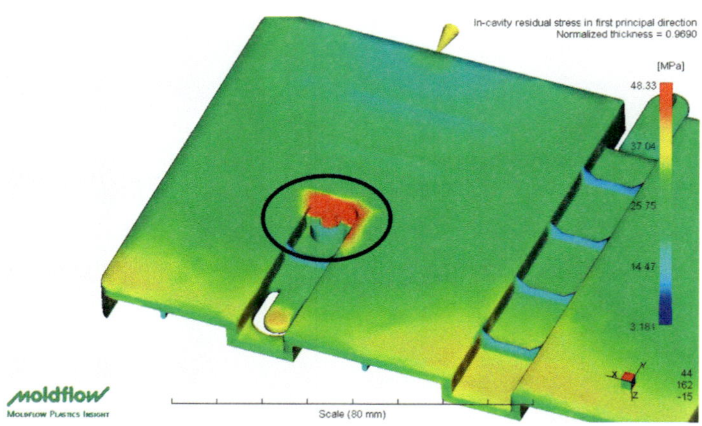

Figure 3.17 Simulation of residual stresses due to localized thick area

Uncontrolled variation in wall thicknesses can lead to catastrophic failures. Figure 3.18 is a turn signal and wiper control lever that actually landed in the driver's hand just in the process of engaging the lever (Figure 3.18 (1))! A 13 mm thickness section was joined to the rest of the part by a 6 mm thickness section (Figure 3.18 (2) and (3)). The result was an obvious crater of a sink mark (Figure 3.18 (4)). This alone should have raised a red flag for the designer that there was trouble lurking ahead (refer to the discussion of thick to thin flow in Section 3.2.6.1.4). Section 3.2.6.1.1.1 talked about the critical thickness of molded plastics. A 13 mm thick section obviously is disastrously above the recommended critical thickness, which is about 3 to 3.5 mm. So the failure here was a combination of extreme thickness and a rapid and large transition in the wall thickness.

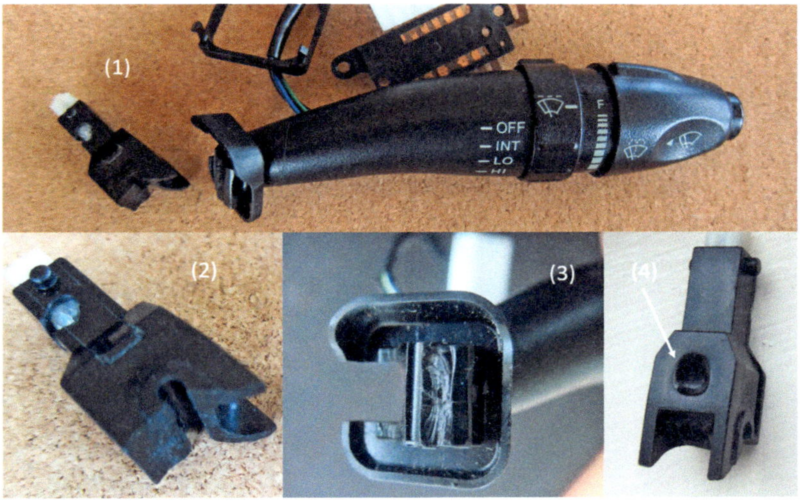

Figure 3.18 Catastrophic failure in automotive turn signal lever

The acceptable variation in the wall thickness in the nominal wall is 25% for amorphous plastics and 15% for semicrystalline ones (see Chapter 2 for the explanation of the two material types). Even the acceptable variation should be gradual. A transition of 1 in 3 has worked well for me leaving no telltale shadows on the outer surface. Thus if a 3 mm wall is being reduced by 0.75 mm (25%), it should decrease gradually over 2.25 mm.

Wall thickness variation may be required to be even less in highly cosmetic areas, as it can result in unsightly "shadows." Figure 3.19 shows one such case. The part is made of ABS. The wall outside the recesses rectangle in the figure is 2.0 mm. The wall thickness in the recessed area is 1.5 mm. Even though the design rule for the variation is followed, it leaves a shadow around the area. The fix for this situation is to reduce the variation or to make the transition more gradual.

Figure 3.19 Shadows left by wall thickness variation

3.2.6.1.5 Rib to Wall Thickness Ratio

Let me start with the acceptable ratio of rib to wall thickness. To avoid the unacceptable appearance of sink marks and improve the physical robustness of a plastic part, the acceptable ratio is 50 to 60 percent of the wall it is attached to.

Figure 3.20 shows the various cross-sectional areas for various rib thicknesses. This is basically the biggest circle that can be inscribed in the cross section and represents the area of the flow channel. The table in Figure 3.20 shows these calculations.

Thus, as an example, if the diameter of the circle that can be inscribed in the nominal wall is 1, the cross-sectional area (CSA) is 0.79 and is a 100 percent of the nominal wall CSA. With a rib thickness of $1/2 \times t$ along with a fillet radius of $1/8 \times t$ it has a cross-sectional area of 0.95 or a 20% increase of the nominal wall CSA. Going back to the allowable wall thickness variation, this is the desired rib thickness. More than this, it can leave anywhere from a visual blemish to a large sink mark with a void in the inside. These sink marks are not merely visual defects. Going back to the discussion in Chapter 2, they introduce a tensile layer on the top. The reason for this is the pull from the warm material in the center as it shrinks. A lot of times, sink marks also leave an internal void if not enough material can be pulled in from the surface when the part is made to cool very rapidly freezing the outside layers (see Figure 3.22). The tensile area leaves the section weak and with a greatly diminished chemical resistance.

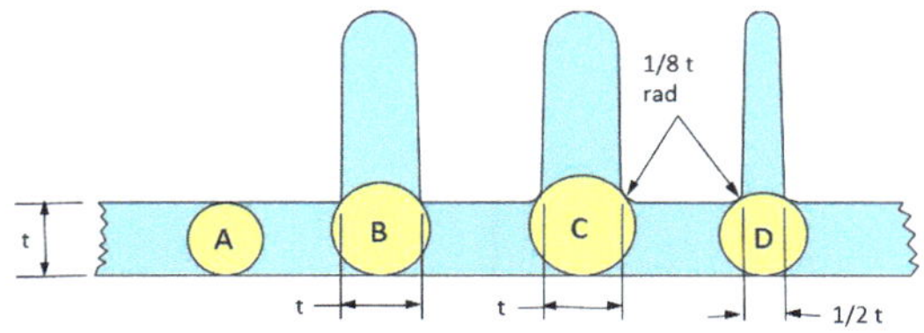

Rib Thickness	Diameter	Cross-Sectional Area (CSA)*	Percent of Nominal Wall CSA
t (A)	1	0.79	100
1t (B)	1.2	1.3	143
1t w/ rad (C)	1.3	1.33	168
0.5t w rad (D)	1.1	0.95	120
		*CSA=π*d^2/4	

Figure 3.20 Cross-sectional areas in various ribs [5]

Figure 3.21 shows a sink mark unusually associated with a thick rib. The top picture shows the very thick ribs protruding from the wall. The bottom view shows the deep sink mark this causes.

Figure 3.21 Sink marks on an incorrectly designed rib

Figure 3.22 shows a common plastic hanger. Normally the thick cross-sectional area is hollowed out by a process called gas injection molding. More information on this process can be found in the following pages. However, in this case, the part was molded by a conventional process. Basically, the "rib" protruding from the "nominal wall" has the same cross-sectional area. To make the molding cycle as short as possible, the mold is operated at very low temperatures. In fact, it is not uncommon to have chilled water run through the mold for parts like these. In the molding section we will talk about the effect on the part properties when the mold is not heated to the required temperature. For the current discussion, what this does is that the outside layers cool down very fast and stop shrinking while the internal layers are still hot and shrinking. This shrinkage pulls the material from the inside causing the void and the failure.

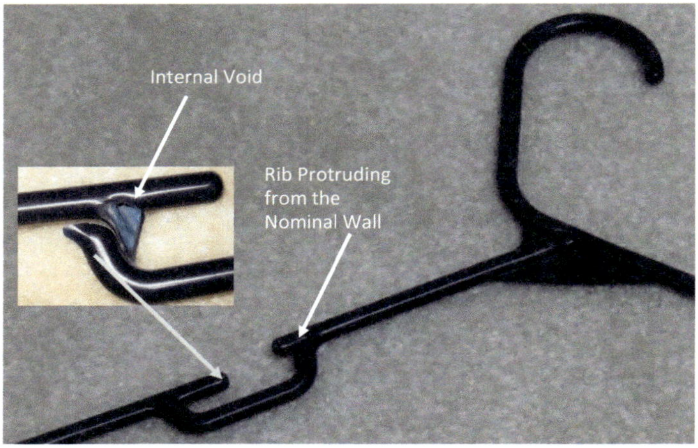

Figure 3.22 Void in a common plastic hanger as the result of a thick rib

Rib Considerations for a Cosmetic Surface

While a ratio of half the wall thickness for ribs along with a minimum fillet radius of 1/8×*t* for structural integrity is commonly acceptable, for highly visual surfaces an even more conservative ratio may be needed to avoid even the very slight visual shadows. Figure 3.23 shows this shadow effect. The top area has a distinct sink mark and the bottom has visually unacceptable shadows. A design checker called DFMPro [12] predicted this condition at the design stage.

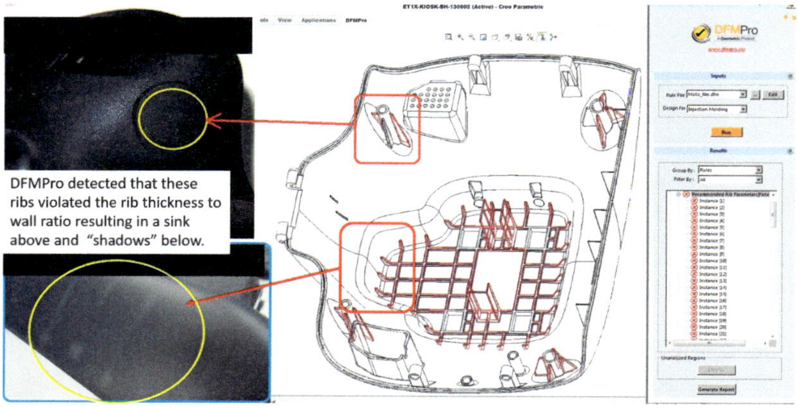

Figure 3.23 Shadows created by thick ribs

3.2.6.1.6 Flow Direction—from Thick to Thin

As was discussed previously, the cross-sectional area of the plastic flow path decreases as the melt flows away from the gate. Now imagine a thick area being filled by a thin area. In most cases, the thick area will not be packed[2] properly due to a very small cross section trying to fill a larger section. Poor packing will result in short shots and internal voids.

The normal way the molder will try to compensate for this is by jacking up the packing pressure and causing higher molded-in stresses in the entire part. More details in the chapter on processing (Chapter 5).

Figure 3.24 is a good example of what can happen when the material is made to flow from a thin region to a thick region. The part is an inexpensive clear polystyrene drink stirrer. The gate is located in an area half as thick as the area it is trying to fill. This results in the thick area not being completely filled and leaving big internal voids. You will also notice some smaller voids in the thin section. These are caused by underpacking combined with the mold running cold as discussed in Figure 3.22.

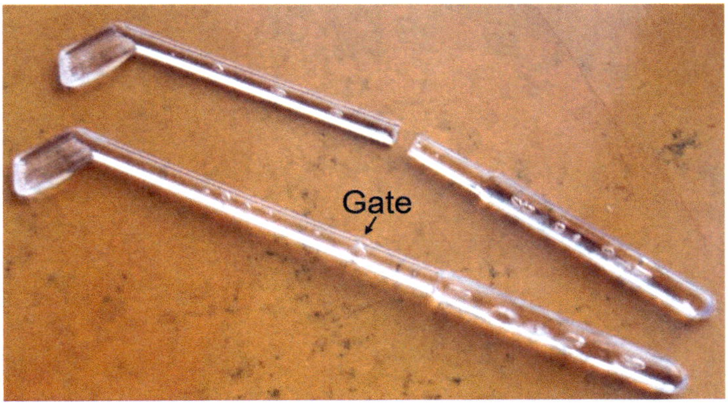

Figure 3.24 Voids caused by thin to thick flow

In a perfect world, every designer would have the freedom to make sure every single design rule was followed. As we all know, this is seldom the case. If the part requirements limit the gate location to a thin area, a feature called a flow enhancer or leader can be employed. It is basically a thicker channel from the gate to the thick location. Please remember that the wall thickness variation limits and corner radii rules apply even to the flow leader! If the wall thickness variation is more than permissible, you will see sink marks on the opposite surface! See Figure 3.25.

---

[2] The normal molding cycles consist of filling, packing, and cooling. Detailed discussion to follow in the chapter on processing (Chapter 5).

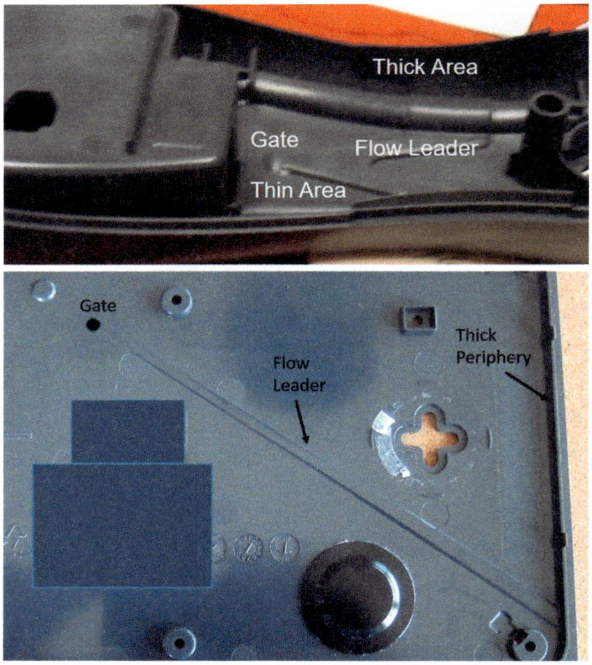

Figure 3.25 Flow leader from the gate in a thin wall leading to proper filling in thicker wall

3.2.6.2 Knit or Weld Lines

Any opening in a plastic part is created by a solid steel core surrounded by a hollow steel structure. The opening between these two is the area that the plastic flows into to create the opening surrounded by the wall. Where the flow fronts meet, a knit or weld line (hereafter referred to as knit line) is created. Figure 3.26 shows the formation of a knit line. The V notch on the surface appears to be a crack. Actually, it was caused by trapped air not allowing the plastic to fill up to the surface.

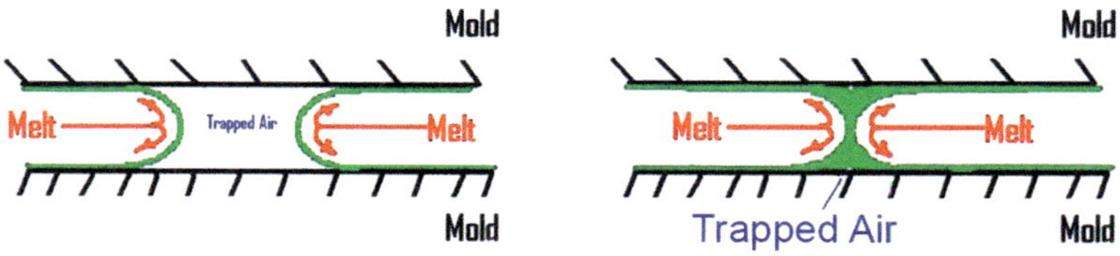

Figure 3.26 Opposing flow fronts creating knit line [8]

Figure 3.27 is an actual cross section of a knit line magnified 400 times. One can see the advancement of the two flow fronts with a tiny notch resulting from trapped air.

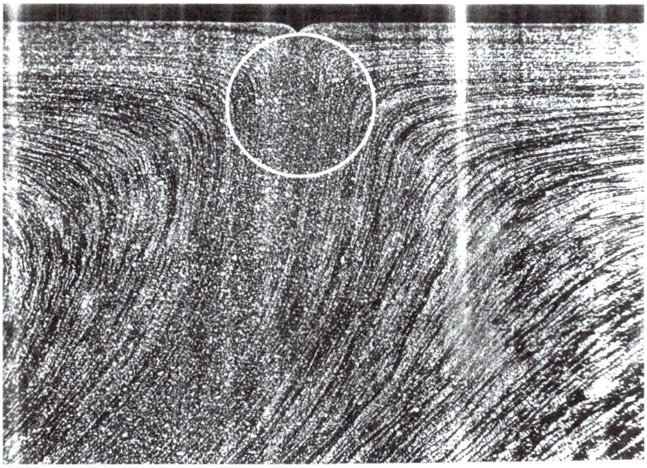

Figure 3.27 400 times magnification of cross-sectional area showing knit line

Knit lines are never as strong as the contiguous wall next to it. While I do not know of any formal studies on the strength of a knit line, the widely held opinion is that a properly processed knit line can have 80-90% of the strength of a non-knit line section. However, poorly filled knit lines can have an actual air gap between the flow fronts—basically having a molded-in crack!

While the occurrence of the knit line may be unpreventable, its location and strength can be optimized by proper gate placement, mold construction, and processing. More discussion on this topic in the tooling and processing section in Chapter 5.

Figure 3.28 shows my own car with a brittle[3] failure at the knit line. This occurred in my very slowly hitting a soft snow embankment—not an uncommon phenomenon driving a car in an area that gets snow in the winter.

The issue is with the placement of the knit line—being exactly where the maximum stress occurs when it hits an object. The clean brittle failure also indicates less than optimum strength from processing and tooling.

[3] More discussion on ductile and brittle failures in the reference chapter (Chapter 9).

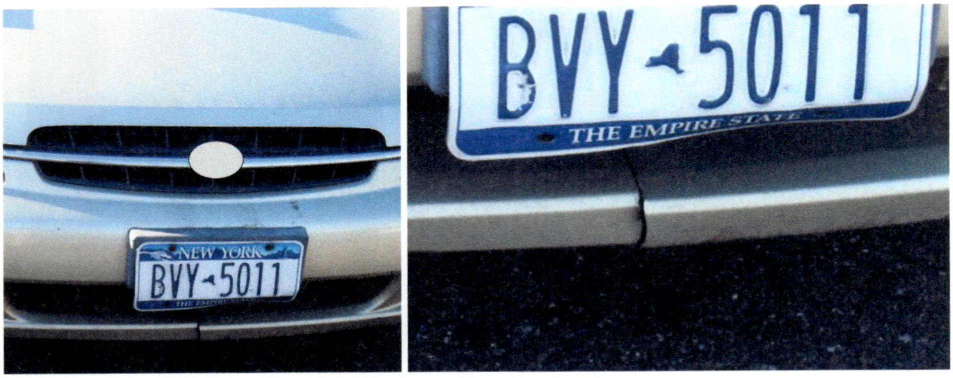

Figure 3.28 Crack in car bumper due to poor location of the knit line

Sometimes it is not possible to place the gate at the ideal location from the knit line point of view based on the configuration of the part. Figure 3.29 shows one such situation. The position of highest impact in this handheld device in a drop is at the 3 o'clock position in the left hand view. The designer does not have a choice in the placement of the gate, which is right at the 9 o'clock position of the opening. With the gate in this position the knit line will be exactly at the likely position (3 o'clock) of impact in a drop.

In a situation like this, a flow leader can be created, which causes an imbalance in the flow to the other end. Looking at the figure one can see the flow leader (basically, a thicker cross section leading from the gate to the right of the opening on the top). This makes the counterclockwise flow faster than clockwise. Thus the knit line is diverted to something more like the 11 o'clock position, as indicated by the view on the right–and away from the position of impact. An additional benefit from moving the knit line is the fact that now it is in a wider area, thus increasing the cross-sectional area.

The image below (in Figure 3.29) shows the Moldflow™ simulation of the flow. Plot 1 is the thickness plot. Plots 2-3 show the melt front pressure as the flow proceeds from the gate to the extremities. Plot 4 shows the final knit line location. Conducting a flow simulation ahead of making the changes will often reduce the trial and error approach, saving a considerable amount of time and money.

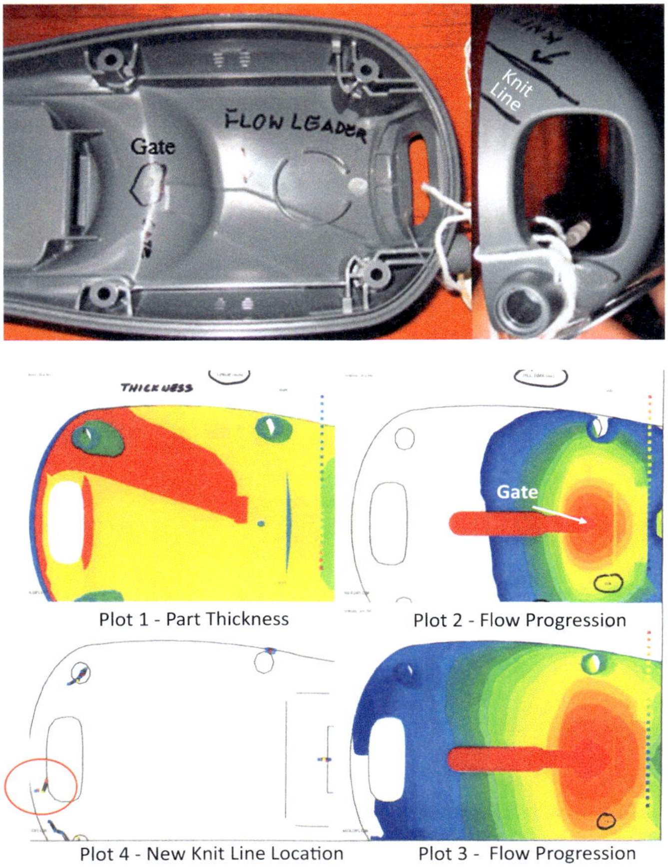

Figure 3.29 Manipulating knit lines

Multiple Gates Result in Multiple Knit Lines

If the wall around a hole is being fed by one gate, it will have one knit line. If created by two gates it will have two knit lines and so forth. Therefore, the minimum number of gates should be used based on the part configuration. A lack of attention to this can create severe performance issues. Figure 3.30 shows a polycarbonate part about 25 mm in diameter with a 2 mm wall. This part was used to hold two separate modules of a handheld electronic device. For whatever reason, the toolmaker chose to have three gates to feed the part. Looking at the figure, the gates are roughly in the 10, 3, and 7 o'clock positions. Therefore, three knit lines are created between the gates—approximately at the 9, 1, and 5 o'clock positions. Three weak areas are thus created in the part. Upon drop, the part failed at the knit lines at the 9 and 1 o'clock positions. This is clearly shown on the flow simulation at the bottom left, which shows the knit line locations and the flow front progression. By closing two gates on the left, two knit lines were eliminated. Even the remaining knit line was more towards the center of the part, further away from the point of highest impact. The flow simulation on the right shows the new flow pattern and knit line. This is a perfect example of a tooling error that can cause a serious performance issue.

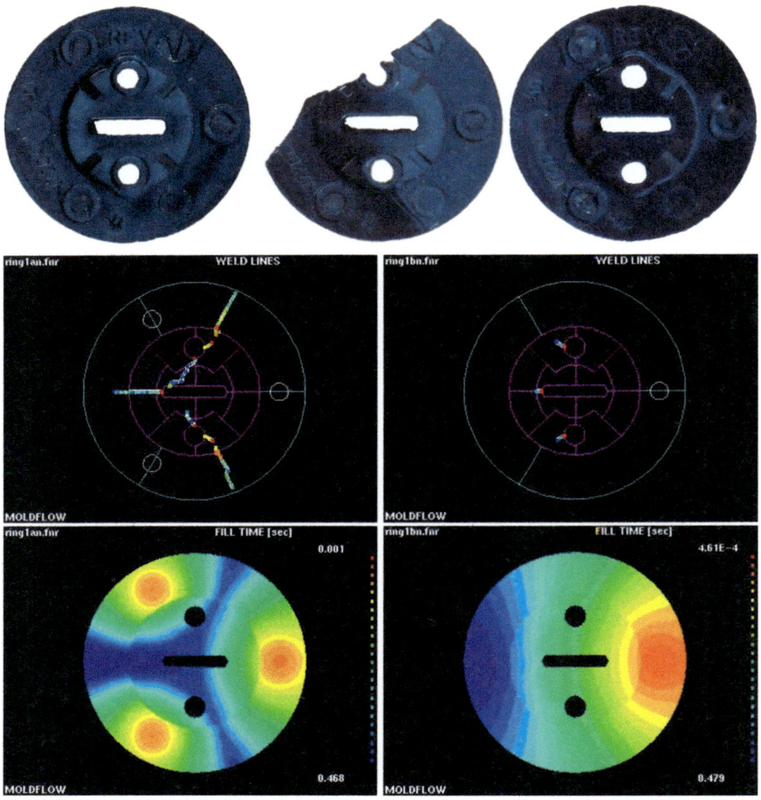

Figure 3.30 Three gates causing three knit lines

3.2.6.3 Boss and Rib Height Considerations

Glenn Beall, who has had a huge positive impact on the design community through his seminars and training aids since the early eighties and who I personally consider my own guru in the area, advocated the rule of thumb of the rib heights and bosses being no longer than three times the diameter or the thickness at the bottom of the hole of the rib. When at all possible this rule should be followed. However, with newer materials, machining technology, and fine tuning the flow through simulation, higher ratios can now be used.

3.2.6.3.1 Bosses

Bosses are created by a steel core pin in the middle of a hollow steel opening. This pin basically acts like a cantilever. Depending on the flow pattern the cantilever can have upwards of 20,000 psi (about 140 MPa) acting on the free end in the worst case. This can bend the core pin resulting in a non-uniform wall thickness around the opening. As time goes on, the core pin can actually break off under fatigue. The sharp corner rule applies here, too. Where possible, the opening on the top of the plastic should be filleted to provide a corner radius on the core pin, thus preventing a stress concentration at the sharp corner.

Figure 3.31 shows a long narrow boss. The length of the hole is about 10 times its diameter at the base. As seen in the bottom left corner, this results in the free end of the pin being deflected, causing the hole to shift to the top left side. As if this was not bad enough in itself, the bending of the pin causes it to be at an angle such that it is not parallel to the direction of draw, thus creating an undercut. When the part is ejected, the ejector actually causes the plastic to be distorted in the process of pushing it out, as shown in the top view by the circle marked in black. (The same part is shown molded in gray plastic.) The pin deflection also causes a heavier section on one side. This heavy cross section violates the 50 % thickness rule and causes a sink mark.

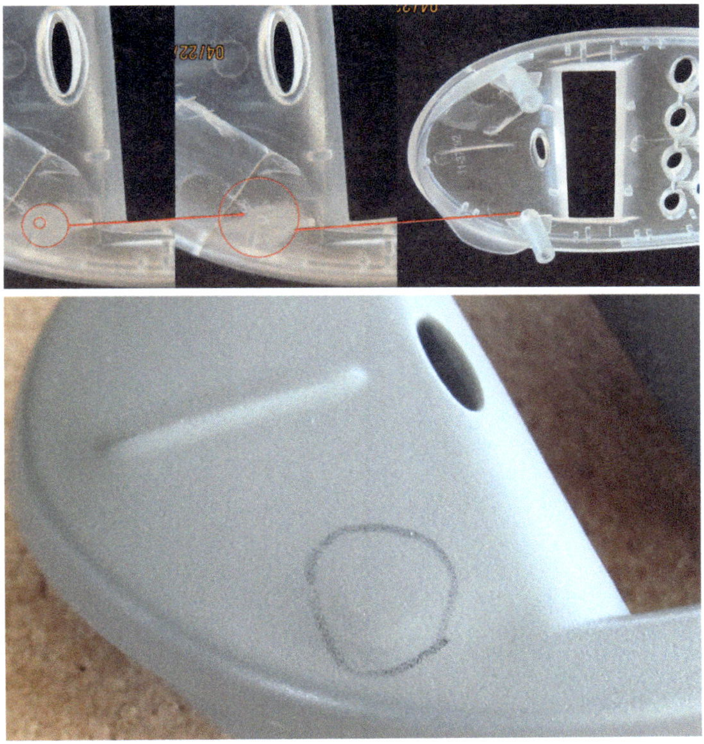

Figure 3.31 Core shift in long thin boss and distortion during ejection

We were talking about a slender core pin in the above example. Now let us look at Figure 3.32. The opening in the part is 250 mm×125 mm×38 mm. This is a massive piece of steel and should not bend, right? Wrong. At the open end of the core block, the bending moment is huge due to the plastic pressure of close to 140 MPa. Enough to bend this column! The result is the wall becoming 1.5 mm on the left and 3.5 mm on the right instead of the 2.5 mm it needed to be.

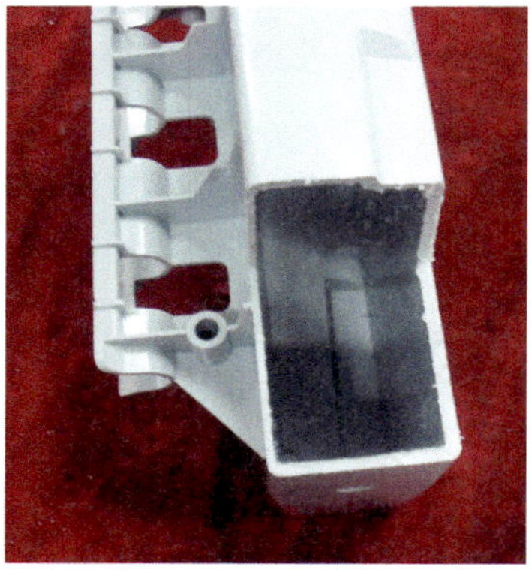

Figure 3.32 Core shift during molding causing non-uniform wall thickness

3.2.6.3.2 Rib Height

The conservative rib height to the base ratio is also about 3. This is due to several factors:

- Polishing of the cavity that forms the rib: The long thin opening is extremely difficult to access and polish.
- Ejection: This is made difficult both by the fact that the cavity is not polished properly and that the ejector pins have to be very thin and not wider than the plastic thickness left after accounting for the draft angle.
- Flow: The pressure required to flow into the thin tall area can leave short shots.
- Mold cooling: The deep thin areas are very difficult to lay a cooling line around.
- Venting: Air and gases get trapped at the end of the flow. Following the ideal gas law:

$$\frac{PV}{T} = Constant \tag{3.2}$$

where P is the pressure, V is the volume, and T is the temperature; therefore, $T \propto PV$.

As the plastic takes up the space previously occupied by the gas/air, its pressure rises much more than the decrease in volume. This can raise the temperature to a point where the plastic actually burns. This is also called dieseling. (As in diesel engines that do not have spark plugs. The mixture of diesel and air is compressed to the point where the mixture ignites.)

Figure 3.10 in the previous pages shows the no flow condition in a deep rib.

Figure 3.33 shows dieseling in a long thin rib that is difficult to vent.

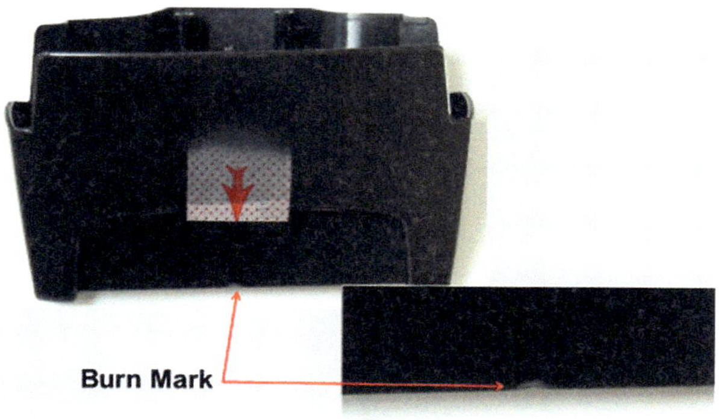

Figure 3.33 Dieseling in long thin rib

■ 3.3 Plastics Structural Calculations

3.3.1 Stress or Strain?

Once again I will repeat the theme: *Plastics are not metals.* Coming from the metals world, we were always concerned about the stress in the cross-sectional area in the case of a directly applied load and the maximum fiber stress in the case of a beam or cantilever being deflected by a load.

The metal property tables gave us dependable values of ultimate tensile or yield stress and tensile or flexural modulus for the exact alloy and temper. We used a conservative factor of safety and reliably calculated the required thickness, width, or length for a safe design.

As discussed in Chapter 2, the physical properties provided by the resin suppliers are only guidelines and can vary widely based on the design, tooling, processing, environmental conditions, and represent a very small temperature range. They can also change dramatically based on the rate of loading. Additionally, the modulus of elasticity E can also change based on the temperature and rate of loading, as seen in Figure 3.36 and Figure 3.49.

Therefore, while the true applied stress and the allowable stress are unreliable, the strain is an exact, reliable parameter. Let us quickly review strain.

Strain is the change of length divided by the original length when a load is applied to a structural member. Thus:

$$\varepsilon = \frac{\Delta l}{l} \tag{3.3}$$

where ε is the strain, Δl is the change in length, and l is the original length.

Most resin suppliers will provide a short and long term allowable strain for their materials.

Looking at the cantilever represented in Figure 3.34 and the labeled dimensions, the following calculations for the strain can be performed without the use of E or the allowable maximum yield stress. Let y be the deflection at the tip of the cantilever under load P:

$$y = \frac{PL^3}{3EI} \tag{3.4}$$

where I, the moment of inertia for a rectangular beam, is:

$$I = \frac{1}{12} bt^3 \tag{3.5}$$

The maximum stress σ_{max} at the outer fiber of the cantilever is:

$$\sigma_{max} = \frac{PL}{Z} \tag{3.6}$$

where Z, the section modulus, is:

$$Z = \frac{bt^2}{6} \tag{3.7}$$

Substituting for I and Z in Eq. (3.6):

$$\sigma_{max} = \left(\frac{3Ey}{L^3} \frac{1}{12} bt^3 \right) L \frac{6}{bt^2} = 1.5 \frac{Eyt}{L^2}$$

The strain ε follows: $\varepsilon = \sigma/E$. Therefore, substituting for σ_{max}:

$$\varepsilon_{max} = 1.5\frac{yt}{L^2} \tag{3.8}$$

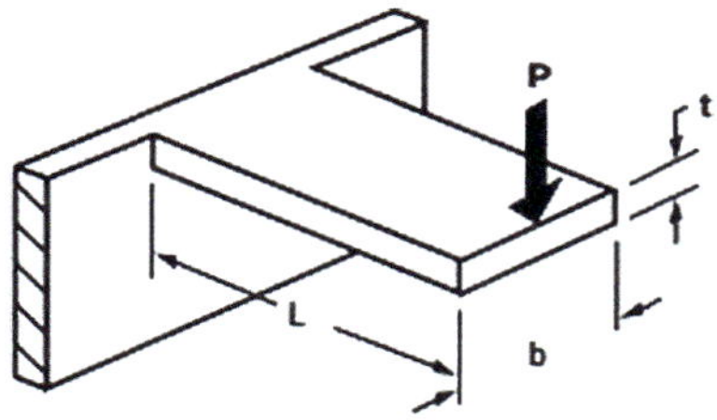

Figure 3.34 Simplified strain calculation on a snap with constant cross section

The next question is how do we get a reliable flexural or elastic modulus if we are working on a beam deflection calculation, since the stress over strain is not a fairly long straight line little changed by normal use temperatures as in the case of metals (see Figure 3.35).

The answer is to use the secant modulus as shown in Figure 3.36 (top). The secant modulus is the local value of stress divided by the strain at any point on the stress strain curve for the plastic *at a given temperature* (see Figure 3.36 bottom).

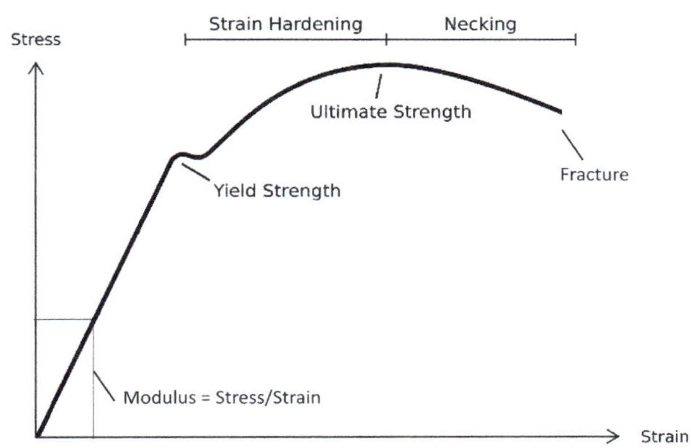

Figure 3.35 Typical stress strain curves for metals [5]

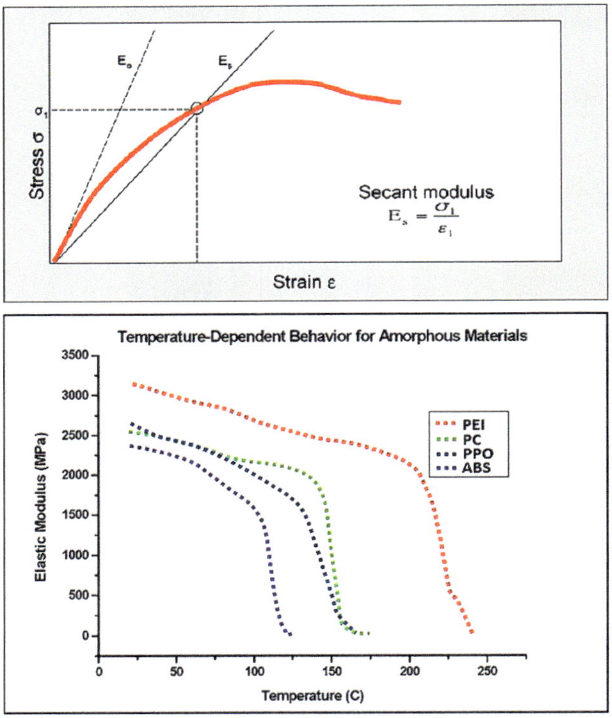

Figure 3.36 Secant modulus, top [5]; temperature-based elastic modulus, bottom

Table 3.1 shows representative permissible *short term* strains for various Covestro resins. Again, be mindful of the temperature of use, since the modulus of plastics drops rapidly with the rise in temperature.

Table 3.1 Representative Permissible Short Term Strains [5]

Apec	High Heat PC	4.0%
Bayblend	PC/ABS	2.5%
Centrex	ASA	1.9%
Durethan	PA cond. dry	6.0% 4.0%
Lustran	ABS	1.8%
Makroblend	Polycarb. Blends	3.5%
Makrolon	PC	4.0%
Triax	PA/ABS	3.4%

3.3.2 Long Term Strains

Repeating the theme *plastics are not metals*, let us now look at long term strains. To prove the point, here is a lighthearted example.

Hang a mass of 100 kg from a steel rod of adequate cross section to withstand the resultant stress. Measure the initial length. Leave a note to your great-grand-children to measure the length in a hundred years. They should see no change in the length.

Do the same with a plastic rod. If you value your life, do not go anywhere under it, even within a few minutes. A combination of creep and stress crazing may make the weight fall within minutes to hours based on the material, temperature, and the environment around it. We are talking about a temperature range of less than 50 °C (e.g., from 20-70 °C) and environments ranging from relatively dry air to humid air, to one with solvent vapors, such as in an automotive garage!

Figure 3.37 shows the stress crazing (see Figure 3.38) curves for a polycarbonate resin from Covestro. The seven curves show the stresses at controlled strain from 0 to 2.5 percent. For the same strain, the stress will reduce as time goes on. This is due to stress relaxation caused by the reorientation of the polymer chains.

The dark dotted line shows when the plastic will start to have stress crazing. Figure 3.38 shows some typical stress crazing. Based on this the plastic may fail between 100 and 10,000 hours or 4 to 400 days under a continuous strain of 2.5%.

Note that these curves are at 23 °C and 40 °C in the clean and controlled lab conditions with 50% relative humidity. Suffice it to say, that in any practical temperature range and humidity, possibly damaging environment, and using any practical factor of safety, the long term strain would be significantly less than 2.5%. In fact, I have seen failures within hours of putting a continuous strain on a plastic component.

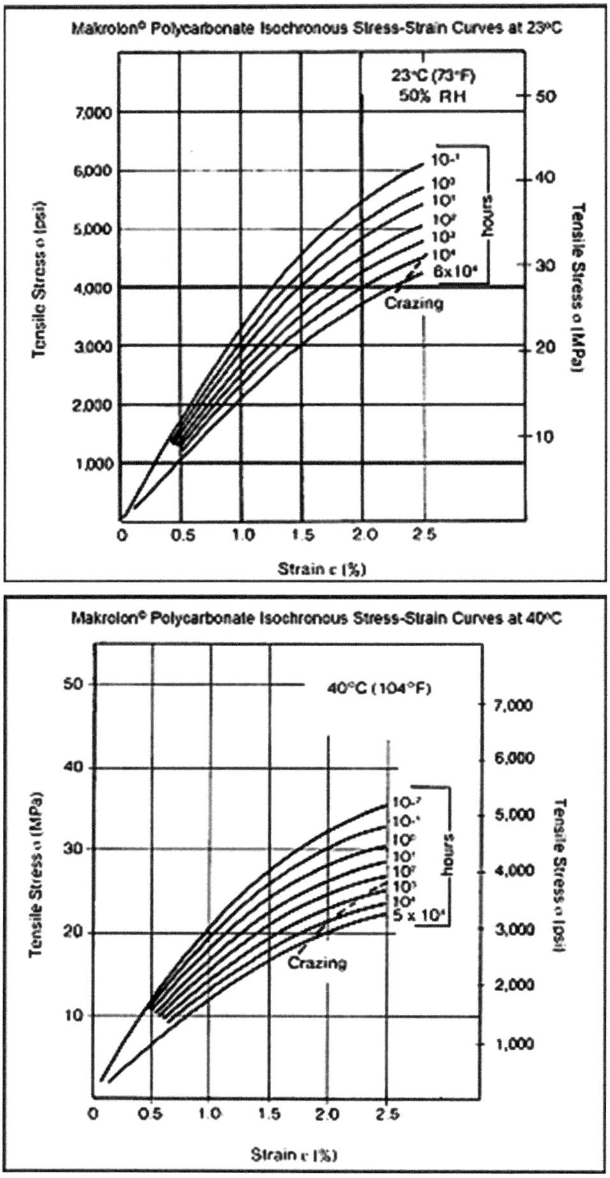

Figure 3.37 Stress crazing curve at 23 °C and 40 °C [5]

Figure 3.38 Stress crazing in a clear polystyrene tumbler

Continued strain can lurk in different corners that the designer may not even be aware of. Here are some common examples.

3.3.2.1 Warpage

Figure 3.39 shows a polycarbonate part that came out warped from the tool. The part was "unwarped" in the process of being assembled with its mating part, thus introducing a continuous strain. In less than a few months cracks appeared in the part just from the sweat on the operators' hands attacking the part under continuous strain.

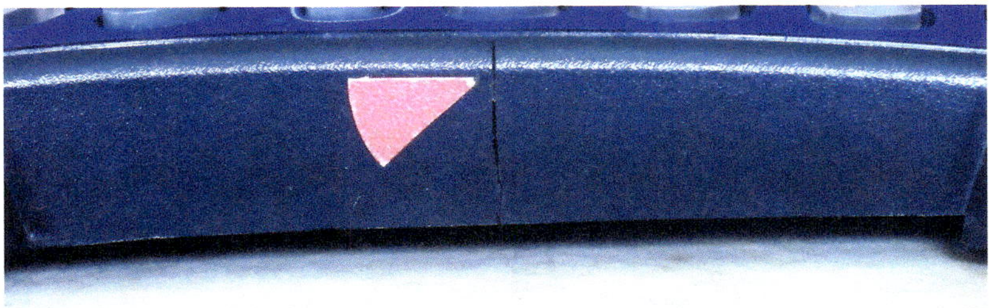

Figure 3.39 Cracks in part "unwarped" when assembled [8]

3.3.2.2 Plastic Screws

There are basically two types of screws used in fastening plastics without the use of threaded metal inserts. They are thread forming and thread cutting. See Figure 3.40: the one on the left is thread forming and the one on the right thread cutting.

The thread forming screw basically forces its way into the plastic boss and distorts the plastic, introducing a continuous hoop strain. The thread cutting screw, on the other hand, has a drill-like cutting edge that cuts its path into the plastic, minimizing continuous strain on the boss.

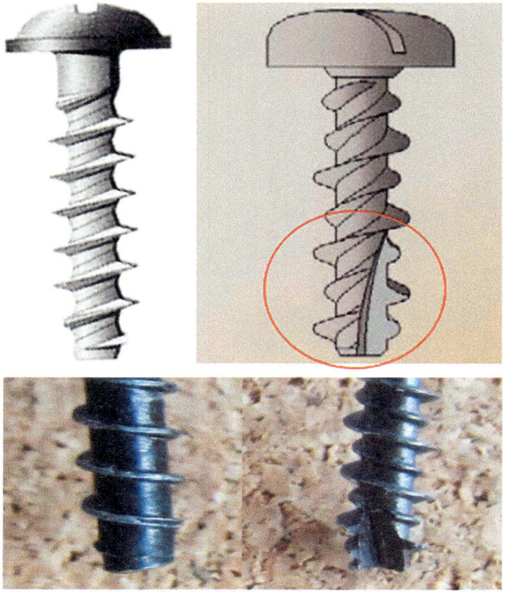

Figure 3.40 Thread forming and cutting screws

Figure 3.41 shows the catastrophic failure of a boss in a PC/ABS part fastened with a thread forming screw. This happened in a matter of weeks after the product was shipped. Changing to a thread cutting screw completely eliminated the problem.

Figure 3.41 Catastrophic failure of PC/ABS part with thread forming screw

3.3.2.3 Force Fit Assemblies

Unlike in metal parts, force fitting a peg into a hole will cause failures. Figure 3.42 shows a portable phone with one cover fastened to the other by force fitting a hole over a stud. In no time at all cracks developed around the hole. I sent the phone back to be replaced by another phone with no cracks. Within a few months cracks developed on the new phone too.

Figure 3.42 Strain crack with force-fitted stud

Flat head screws work wonderfully well for metals and wood.

One may design a conical opening for the flat head in a plastic part. However, the seating torque required to prevent the screw from coming loose introduces a continuous hoop strain in the opening. Figure 3.43 and Figure 3.44 show the failures in low and high modulus materials respectively. The part in Figure 3.43 is made out of high density polyethylene with an elasticity modulus of approximately 400-1,000 MPa. The failure is thus in the ductile mode[4] and shows the white stress marks around the hole. Figure 3.44 shows the failure in an acrylic sheet. This is a much stiffer material and the elasticity modulus is approximately 2,800 MPa. This leads to a catastrophic brittle[4] failure.

---

[4] See discussion on brittle and ductile failure starting in the reference chapter (Chapter 9).

Figure 3.43 White stress cracking marks with flat head screw in a pclyethylene part

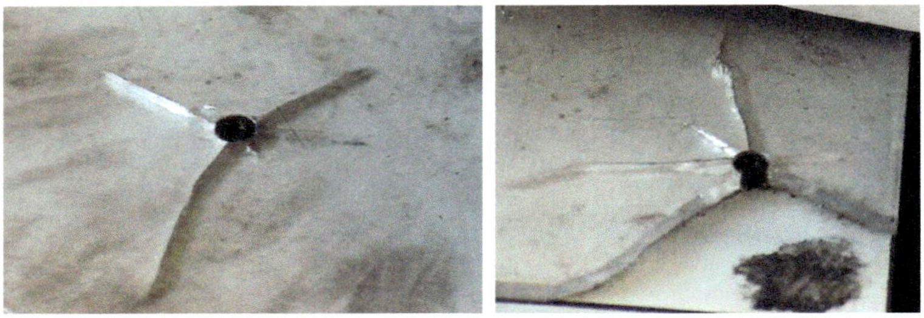

Figure 3.44 Catastrophic failure with use of flat head screw in acrylic sheet

3.3.2.4 Tolerance Stack Up Issues

This kind of failure is very common and occurs due to continuous strain attributable to tolerance stack up issues. Figure 3.45 shows the tolerance analysis of the top and bottom covers of a power supply made out of polycarbonate. When assembled, surface A should rest against surface B and surface C against D. Under the theoretical nominal conditions all of them come together when assembled. However, in the worst case A and B have a gap of 0.030. This does not even take into account additional possible gaps due to part warpage.

The continuous strain, therefore, is approximately 0.030/0.500 = 6% because of the strain being borne solely by the 0.500 length above the four gussets on the long boss. Also to be noted is the fact that these components can easily be subjected to a continuous temperature over 60 °C in use.

Figure 3.46 shows the actual failure of the long boss under these conditions.

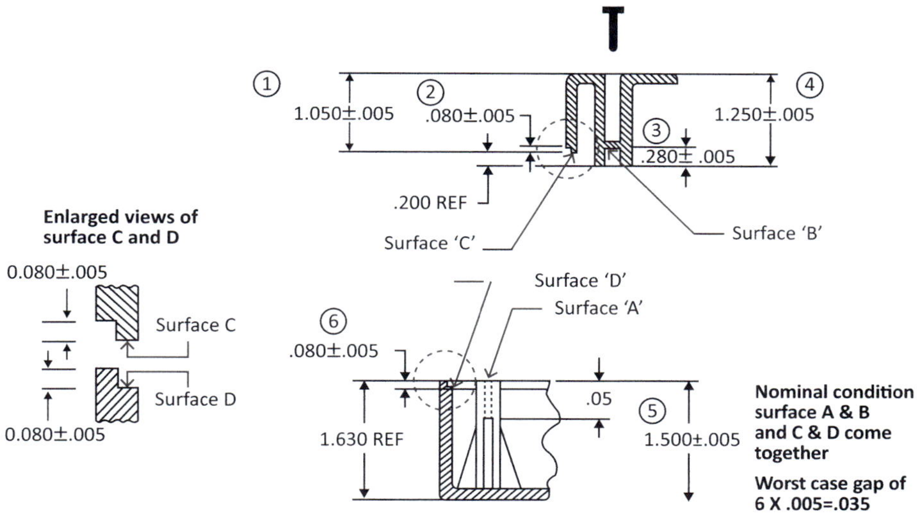

Figure 3.45 Analysis of worst case gap

Figure 3.46 Boss failure due to tolerance stack up issue

It is not uncommon for enclosures to have a set of bosses that need to come together along with another pair of surfaces when assembled. Therefore, the surface of the boss will have a gap with its mating surface if toleranced properly, leading to a continuous strain. In order to prevent this strain on the bosses, a set of "crush cones" can be added to the surface of the boss. These cones "crush" themselves to fill up the gap, preventing a continuous strain on the bosses. See Figure 3.47 for illustration. Figure 3.48 shows the actual parts that had no issues of long term strains.

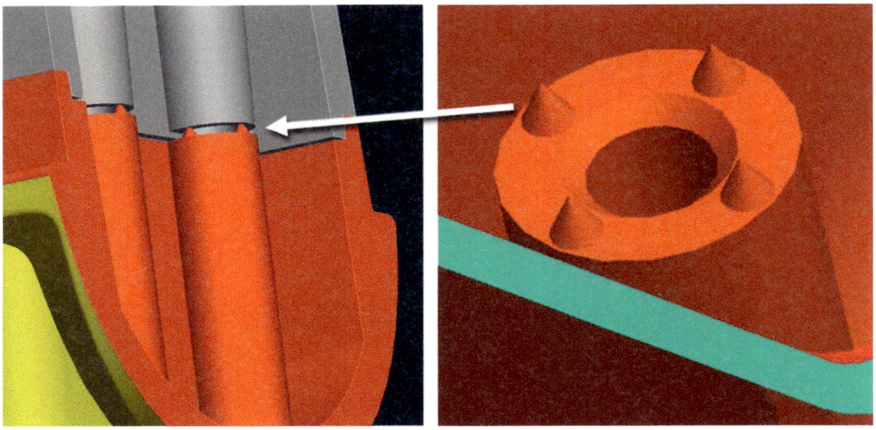

Figure 3.47 Insert crush cones to take up tolerance stack issues

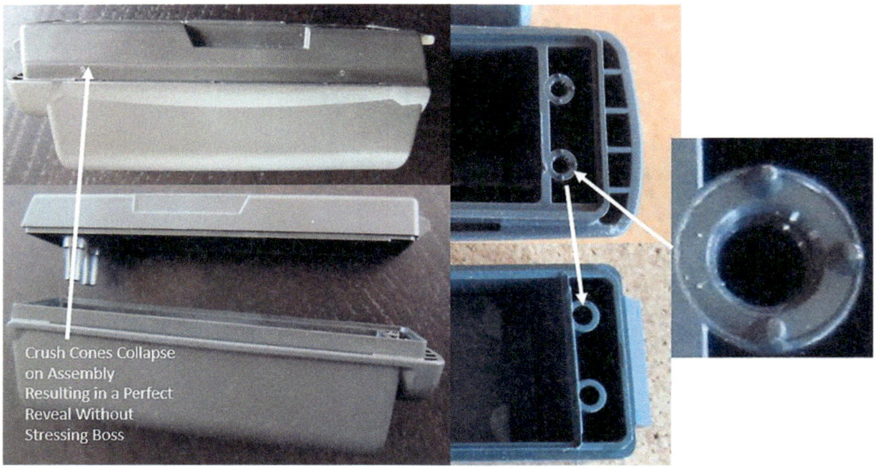

Figure 3.48 Actual part with crush cones

3.3.3 Change in Physical Properties with Rate of Loading

One final word of caution. The modulus of elasticity changes with the rate of loading. Thus, the modulus of a plastic component is much stiffer at a high rate of loading (such as in a drop) than with a stationary load applied to it.

Those who are designing parts to survive drops from various heights or other high rates of impact need to pay special attention to the strain rates and the temperature at the application of stress. Figure 3.49 shows this relationship in general.

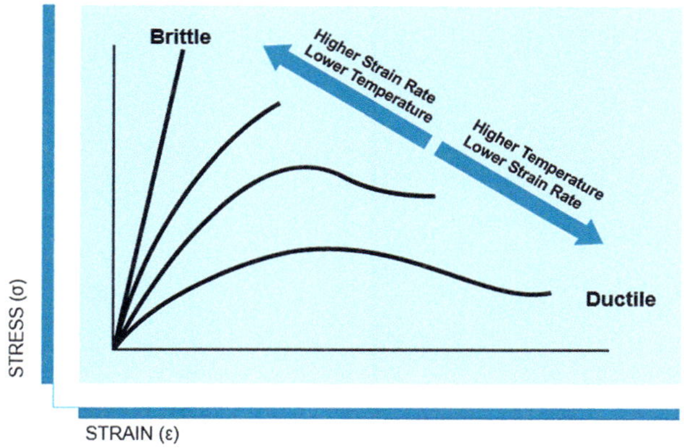

Figure 3.49 Brittle and ductile behaviors of plastics based on temperature and rate of loading [5]

3.3.4 Metal Inserts

It is not uncommon to see metal inserts molded into plastic parts. There is a fundamental issue with this. Look at Table 3.2, which shows the coefficient of thermal expansion for various common materials. Thus, one can see that the coefficient of polycarbonate is about 5 times that of steel.

Table 3.2 Coefficient of Thermal Expansion of Common Materials [5]

Material	in/in/°Fx10^{-5} (mm/mm/°Cx10^{-5})
Glass	0.5 (0.9)
Steel	0.8 (1.4)
Composite RIM	0.8 (1.4)
Brass	1.0 (1.8)
Aluminum	1.3 (2.3)
Nylon GF*	1.3 (2.3)
Polyester GF*	1.4 (2.5)
PPS GF*	1.5 (2.7)
Polycarbonate GF*	1.7 (3.0)
ABS GF*	1.7 (3.0)
Polypropylene GF*	1.8 (3.2)
Acetal GF*	2.5 (4.5)
Acrylic	3.8 (6.8)
Polycarbonate	3.9 (7.0)

Now let us look at what happens when the part comes out of the mold. The plastic needs to shrink five times more than the steel that is preventing it from doing so. The result is a continuous strain on the plastic that can make it crack. Even if it does not crack, its load carrying capacity and chemical resistance are substantially diminished. Figure 3.50 shows the simulation of the strain because of the unequal shrinkage. The dark blue area in the middle is a steel plate completely encapsulated by PC/ABS. You can see that the continuous strain in the corner is approaching over 2.5 %. The part in question ultimately cracked in a matter of weeks.

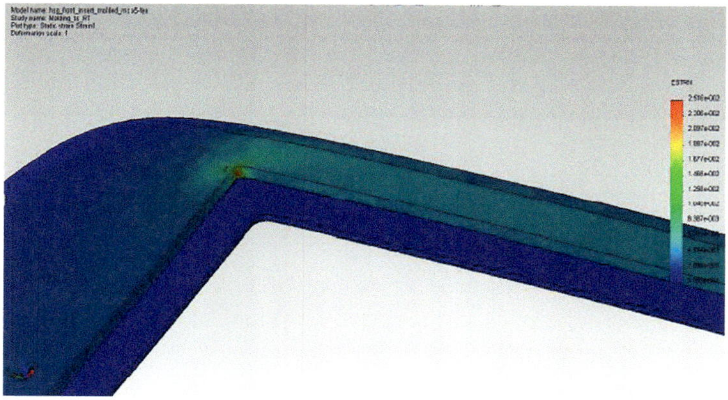

Figure 3.50 Simulation of strain with molded-in steel plate shown in dark blue [8]

Molded-in inserts are another bad idea. Looking at Figure 3.51 (top), one can see the catastrophic failure that it led to with the application of common household cleaning agents around the molded inserts in a polycarbonate part.

The bottom picture in Figure 3.51 is the hub of a radiator fan where the metal hub is overmolded with glass-filled nylon. Since the metal part is larger than the plastic, this is also called "outsert" molding. Notice the cracks in the plastic along the periphery of the hub. The plastic is not allowed to shrink to its desired length by the metal, resulting in these cracks and the ultimate catastrophic failure of the assembly.

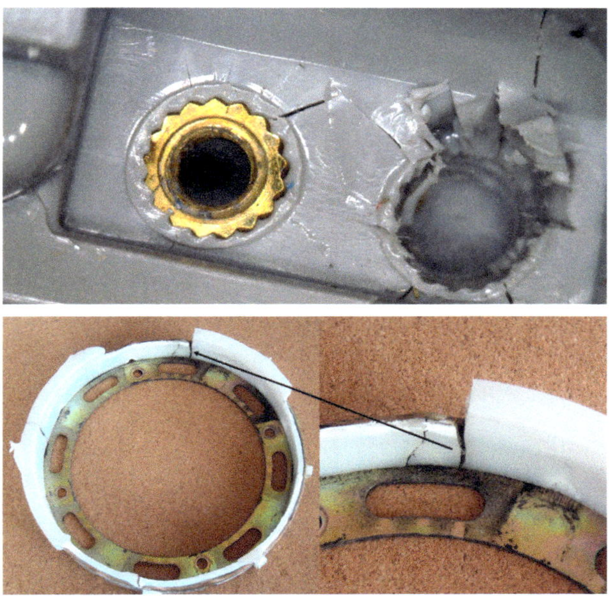

Figure 3.51 Catastrophic failure of molded-in insert with the application of common cleaning agent [8] and "outsert" molded fan blade assembly

There is some argument that lower modulus materials such as HDPE can withstand molded-in inserts better than their stiffer counterparts, such as polycarbonate, acrylic, and polystyrene. In my view, even these plastics may also ultimately fail, given that their coefficient of thermal expansion is around 6 to 11×10^{-5} in/in/°C or almost up to four times that of PC, depending on the flow direction. Figure 3.52 shows a molded-in steel blade in a common letter opener. The material is HDPE. Even this failed, as the material was held in a constant state of strain.

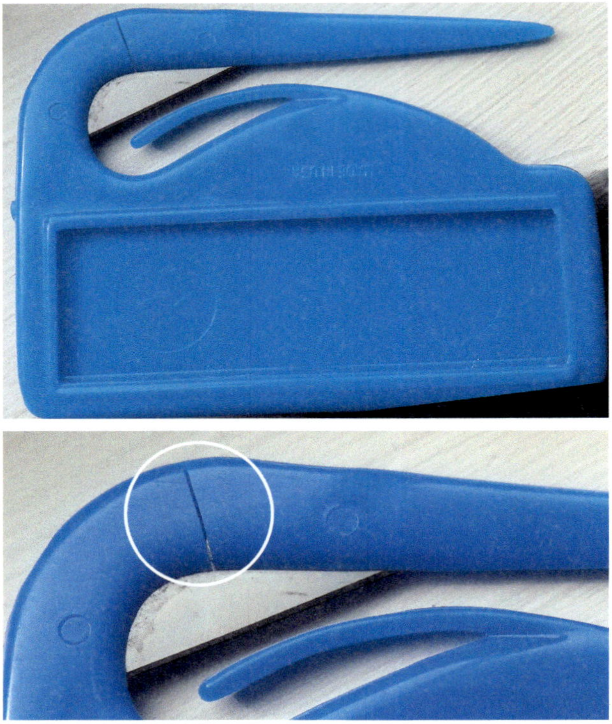

Figure 3.52 Letter opener insert-molded blade failure

Here is another example. Figure 3.53 shows the handle of a pair of scissors. The handle consists of polypropylene molded around the scissor blade. The dotted white line indicates the outline of the blade. The area to the left of the crack had the normal post-molding shrinkage. The area to the right of the crack was not allowed to be shrunk. This caused a very high stress at the junction leading to the ultimate failure shown on the right.

Here is a word of caution about rigidly fastening metal parts to plastic parts. Figure 3.54 shows a plastic metal keyboard assembly that I had the dubious honor of designing. The assembly consisted of a conventional keyboard steel chassis to which were assembled the keys and the electronics. This was about 17 inches long. The top and bottom ABS covers were fastened to each other with the chassis being rigidly sandwiched between the two screws, each fastened on the two ends. Thus the steel and ABS components could not move against each other.

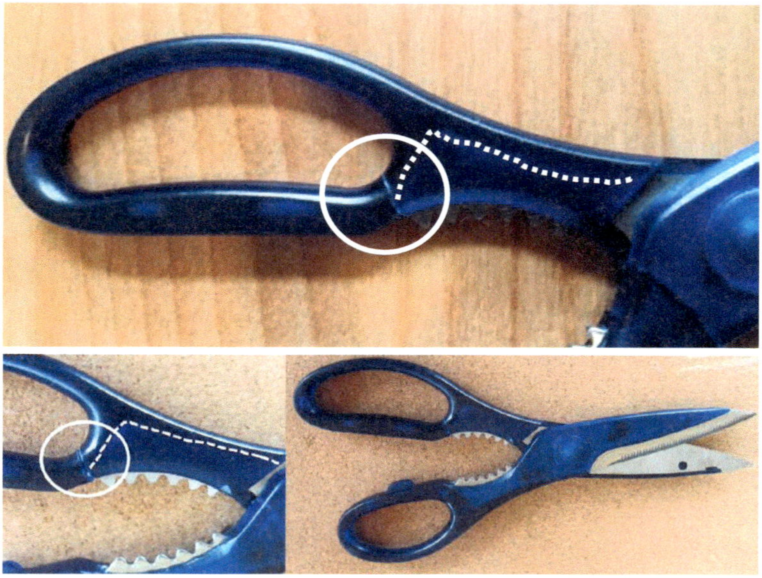

Figure 3.53 Broken handle of scissor blade overmolded with plastic

The product needed a shipping test that included a continued long exposure to the extremes of −20 °C and 65 °C, alternating from one cycle to the other for 72 hours. At the end of the test the assembly came out with both the top and bottom covers severely bowed causing a big gap in the middle.

Let us do the math.

The covers were made of FR (flame retardant) ABS. The assembly was conducted at about 20 °C. The highest temperature was 65 °C. Therefore, the assembly was subjected to a thermal expansion of 45 °C or 113 °F. Using Table 3.2, the steel expanded by 0.010 inches. The coefficient of thermal expansion for ABS is about 4.0×10^{-5} in/in/°C and therefore it expanded by about 0.077 inches. This caused the top and bottom cover to bow. The heat deflection temperature of flame retardant ABS was about 70 °C at 266 psi. Many hours of exposure close to the heat deflection temperature caused stress relaxation and the covers took a permanent set.

Just to complete the story, the solution that I came up with was to put a nylon washer at both sides of the steel plate. With its low coefficient of friction against steel, the covers were able to move in and out relative to the steel during the heating cycle. Problem solved!

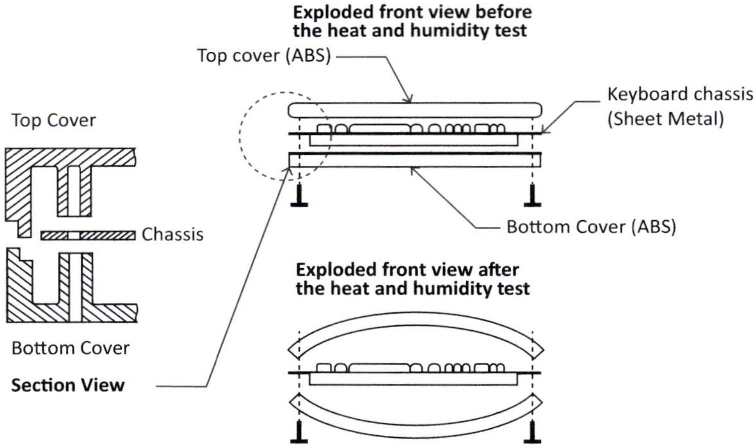

Figure 3.54 Schematic diagram of metal plastic keyboard assembly

Overmolding a soft plastic over a hard plastic is very common these days. There are two common methods to do this. The first one is by using a two-shot machine, which basically has two barrels–one for the hard plastic and the other for the soft plastic. One material is overmolded over the other in a rotating mold. For the purpose of the discussion here, both materials are molded when they are hot and cooled down together when ejected from the mold. This leads to minimum distortion of any one of them.

The other process is called sequential molding, where the hard plastic (called the substrate) is molded in a separate machine. The substrate is molded as an insert for the overmolding in the second machine. Let us take the example of an ABS being overmolded with a TPU (thermoplastic polyurethane) elastomer–a very common combination. Here are the recommended molding conditions and thermal properties for a TPU Texin 285 from Covestro:

- Melt temperature: 200 °C
- Mold temperature: 40 °C
- Coefficient of thermal expansion (TPU): 13.1×10^{-5} mm/mm/°C
- Coefficient of thermal expansion (ABS): 8.8×10^{-5} mm/mm/°C

The ABS part is put into the mold at room temperature and it is at a much lower temperature than the TPU when ejected. Ultimately, both materials cool down to room temperature. As a result, the ABS, with its lower coefficient of thermal expansion and lower cooling range, shrinks almost only about a fifth of the TPU.

Figure 3.55 shows a test specimen used for testing the adhesion of an elastomer with a hard plastic. In this case the materials are TPU and ABS. The bar is about 5 inches long.

The TPU shrinks about five times more than the ABS and causes the ABS to be bowed in. In an actual part this bowing in can cause the distortion and ultimate failure through environmental stress cracking. It can also make the adhesion much weaker as common solvents, such as human sweat, can penetrate the interface between the materials.

Figure 3.56 shows the severe separation of the overmold from the substrate because of human sweat penetration.

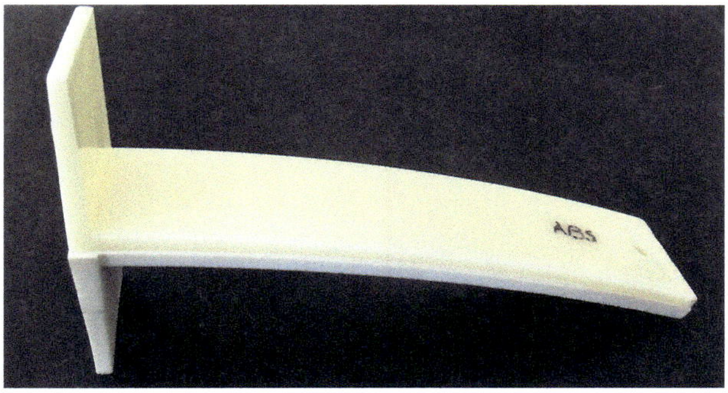

Figure 3.55 Warpage of two plastic parts overmolded on each other

Figure 3.56 TPU peeling away from substrate

3.3.5 Draft

Figure 3.57 shows a simplified view of the draft on a part. Just to establish a common terminology, angle "a" is on the cavity side created by a hollow cavity; angle "b" is on the core side created by a solid steel core. As the plastic cools, surface "A" shrinks away from the steel and surface "B" shrinks onto the steel surface. Theoretically, surface "A" does not need a draft. However, if it did not, the draft on surface "B" would create an uneven wall with the top being much thicker than the bottom. As we discussed before, a uniform wall thickness is one of the most important requirements for a plastic part.

Additionally, we need to make sure that the part does not stay on the cavity side during the mold opening. If it does, the part will be very difficult to remove, as there is no ejection mechanism on the cavity side.

Finally, in a lot of instances, surface "A" may have a texture. See Table 3.3.

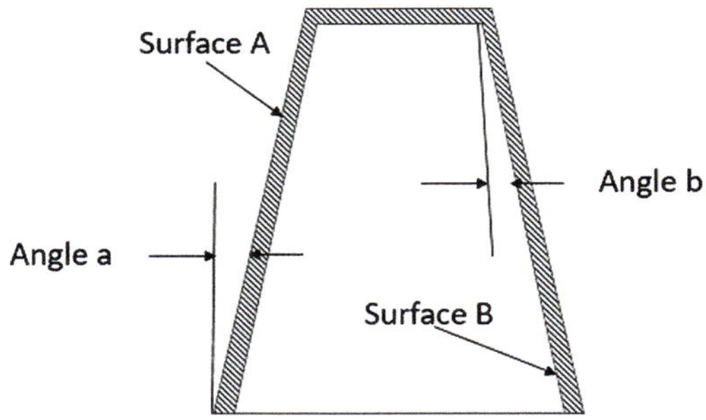

Figure 3.57 Simplified draft requirements (internal corners not radiused for clarity)

In the process of cooling the plastic shrinks five times or more than the steel it surrounds. Therefore, prior to ejection the plastic really force-fits itself over the metal core. In the absence of a reasonable draft, even in a perfectly polished core the part is very difficult to eject. Because of this,

- the ejector pins dig themselves in the part and/or cause local distortion,
- the part gets severely deformed as one side may push out more easily than the other, and
- unappealing scuff marks may result.

To combat this, the molder may run the mold colder with a longer cycle time so the part is stiffer at ejection. This, in turn, causes the part to have additional and unnecessary molded-in stresses that reduce its robustness. As discussed earlier, this may also substantially increase the cost of the part.

Additionally, a part may get stuck in the mold. Worse still, a part of the part may break-off on the core. If this happens there may be a considerable downtime while the full or partial part is removed. It is not uncommon to have to dismount the mold and take it to the tool shop to have the stuck part machined out.

This downtime has the obvious effect of reduced throughput. Additionally, the process of bringing the machine back on cycle may lead to the production of tens, if not hundreds, of parts that need to be scrapped. Once on cycle, the part has to go through re-inspection. Even with the re-inspection, defective parts may make their way into good parts.

As W. Edwards Deming said, "Variation is the enemy of quality." Every time a machine is restarted, it introduces an unnecessary variation.

The larger the draft angle, the easier the molding. However, there is a practical limit to how big the draft can be due to the functional and aesthetic needs. There is general agreement that a minimum of 1/2 degree of draft is acceptable. If the part is textured, additional draft is required.

Mold-Tech® is one of the leaders in providing textures on molded parts. Figure 3.58 and Table 3.3 specify the minimum draft requirements for these textures. This is because the texture produces minor pits in the steel, creating undercuts. If the plastic does not have the required draft, it will result in scuff marks on the textured surface. Additionally, the parts can get stuck on the cavity side, causing serious downtime as discussed above.

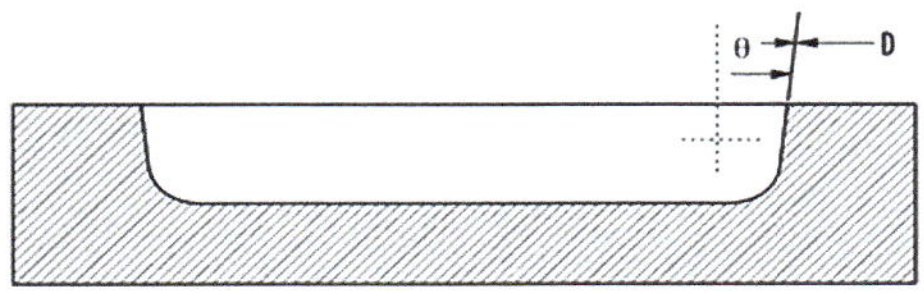

Figure 3.58 Draft requirement on textured cavities [11]

Table 3.3 Draft Angle Requirements Based on Figure 3.58 [11]

TEXTURING NO.	DEPTH (D)	DRAFT ANGLE (θ)
MT-11000	0.0004"	1°
MT-11010	0.001"	1.5°
MT-11020	0.0015"	2.5°
MT-11030	0.002"	3°
MT-11040	0.003"	4.5°
MT-11050	0.0045"	6.5°

3.3.6 Gate Location and Effect of Location

Gates serve as the opening in the steel wall through which plastic enters the cavity. In the process of filling and packing the furthest area of the part the gate area gets overpacked. Containing the hottest temperature melt, it also has a lot of thermal energy left when the part is ejected. For all of these reasons it has a larger proportion of layers under tensile stress as compared to the layers under compressive stress. It is therefore a very weak area of the part and should not be located in an area that sees high stress.

Try this at home. Take an everyday injection molded clear polystyrene cup. Now locate the gate at the bottom. Take a pen or pencil and tap anywhere on the bottom surface away from the gate. In every case, the cracks will emanate from the gate, indicating the fact that the gate is where the crack starts. See Figure 3.59.

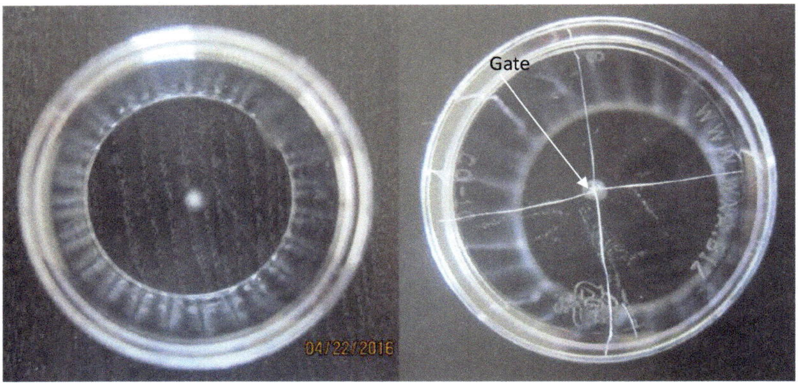

Figure 3.59 Cracks emanating from gate in simple polystyrene cup

The birefringence image (Figure 3.60) shows that the highest stress area is the gate location. Birefringence is created by sandwiching the part between two polarized sheets. It is an effective method to see the stress patterns in a clear part. More discussion on birefringence in the reference chapter (Chapter 9).

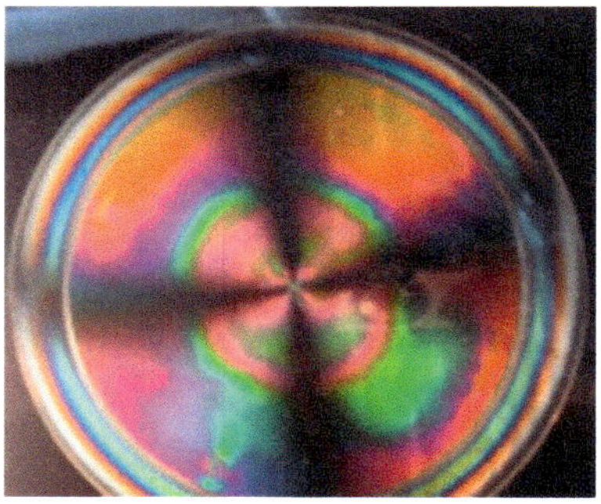

Figure 3.60 Birefringence image of the gate location

Figure 3.61 shows the stirrer used in earlier discussions. No matter where you bend the part, it breaks at the gate shown by the circle.

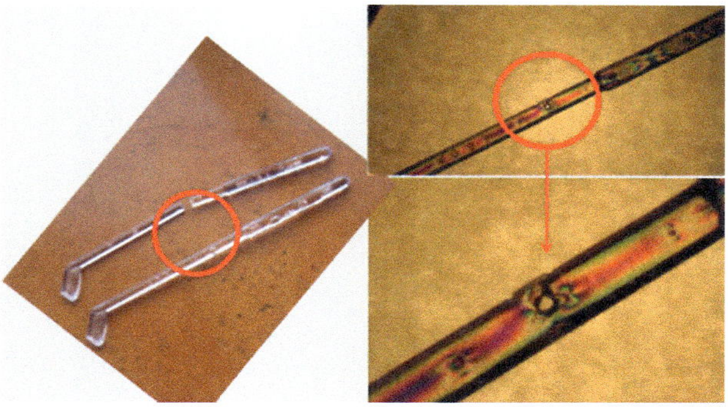

Figure 3.61 Gate location on polystyrene drink stirrer along with the birefringence image

With the above discussion in mind, the gate location should be based both on balanced flow and functional requirements of the part. It is therefore reasonable that the designer be involved in deciding the location of the gate and that this is not just left to the flow simulation or tooling personnel.

3.3.7 Orientation

Figure 3.62 (top) shows the crack pattern when a polystyrene cold cup is crushed. You can clearly see the very straight and parallel lines that the cracks result in. This is because of the orientation of the polymer chains.

In normal processing, with the molds being heated to the optimum temperature, the polymer chains are allowed to go back to their normal random state before ejection. However, in the case of the cup, to reduce the cost, the cycle time is extremely low. To help reduce the cycle time, the mold is run very cold. The material freezes almost instantaneously upon injection and the chains do not have the required temperature or time to go back to the random entanglement state. As pointed out in Chapter 2, the chains are lined up so the chemical bond is only in the direction of the flow. In the cross direction there are only the weak secondary bonds. Thus, on impact the cracks develop across the secondary bonds. In Figure 3.62 (bottom) there is the same cup, in which water was boiled for a few seconds in a microwave oven. The chains have now reoriented themselves into their natural position and the cup itself is much tougher.

Polystyrene is an amorphous material. In the case of a semicrystalline material, running the mold at the optimum temperature is even more critical. If the mold is run cold, the crystalline areas are not allowed to develop during the molding cycle. The part is therefore very weak and will distort over time as the crystalline areas develop.

Figure 3.62 Orientation in common polystyrene cup

If life gives you lemons, make lemonade.

The fact that chains get oriented under certain processing conditions can be taken advantage of. The common polypropylene bags are oriented such that chains run in the direction of the handle. Thus, it takes very little effort to tear the bag at the right angle to the handle direction and it is extremely difficult to do so in the long direction of the handle. Figure 3.63 illustrates this.

Living hinges are another common example of taking advantage of orientation and the resultant superior properties in the length direction of the chains. Figure 3.64 shows a simplified illustration of a living hinge. The material is made to flow from one thick section to the other through the thin one. Ideally, the thin section is cooled rapidly by putting cooling lines directly above and below it. This freezes the chains across the thin section, giving it an enormous fatigue life. More information on designing living hinges in the reference chapter (Chapter 9).

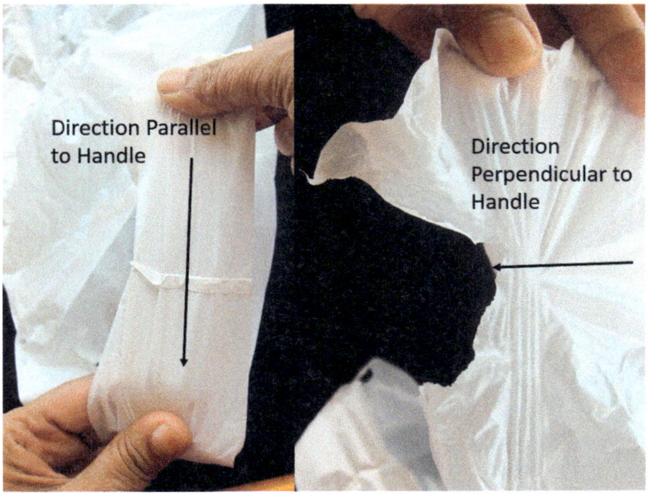

Figure 3.63 Orientation effect in common polypropylene shopping bags

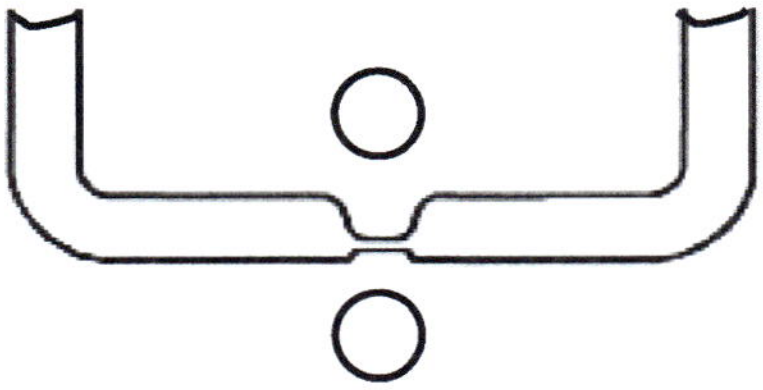

Figure 3.64 Simplified living hinge illustration

Figure 3.65 shows a common a dental floss container with a very efficient living hinge.

Figure 3.65 Common dental floss container with living hinge

3.3.8 Wear of Plastics

Care has to be taken when designing plastic members that slide against each other during use.

When two plastics rub against each other, the materials from the surface are gradually removed due to adhesion or abrasion. Adhesive wear takes place when very small pieces of the materials are removed by the rubbing surface. This results in a fine dust. When a harder surface digs up a softer surface, abrasion takes place and this results in grooves or scratches on the surface.

The button in Figure 3.66 was made of PC and was supposed to rub against the mating surface also made of PC. Most amorphous materials have a very high surface energy and therefore good adhesion. Just a few hundred cycles caused the button to get sticky due to adhesive wear. The button was changed to acetal and the combination worked as desired.

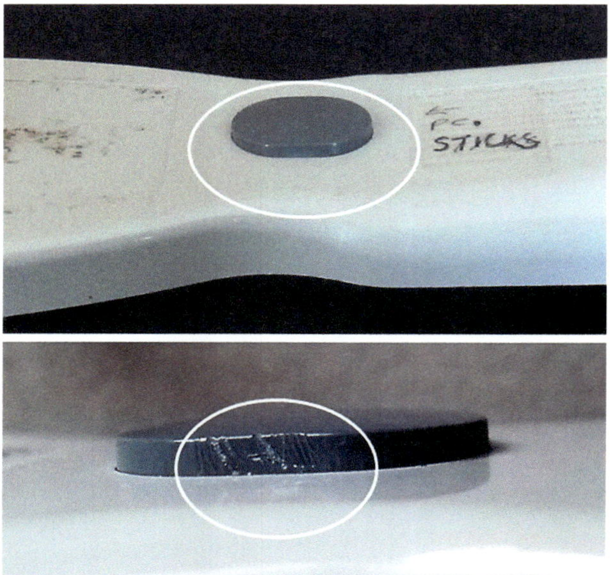

Figure 3.66 Trigger button failure: the circled area shows vertical wear scratches on the button

3.4 Thin Walls, Structural Foam, and Gas Assisted Molding

Let us now examine three special kinds of moldings. Some rules of conventional injection molding in terms of wall and rib thickness are different. Other rules, such as internal corner radii and draft, still apply. Let us look at them individually.

3.4.1 Thin Wall Molding

Thin walls refer to parts with nominal wall thickness of 1 mm or less for small parts, such as cell phones, and 2 mm for larger ones, such as automotive parts. This is brought about by the need for thinner profiled, lighter products such as cell phones, which are fighting for the reduction of every gram of weight and every millimeter of the thickness of the device.

Figure 3.67 is a typical relative viscosity-shear rate curve for injection molding. More discussion on this will follow in the chapter on processing (Chapter 5). However, a simple explanation follows, for the purpose of understanding thin wall molding. The X axis represents the shear rate or the reciprocal of the fill time (injection phase). The Y axis represents the relative viscosity of the melt. (Note that relative viscosity here does not refer to the relative viscosity as in the viscosity of the solution divided by that of the solvent used in viscosity studies. This is a simple dimensionless measure of the relative ease of flow *within this particular study*.) The ease with which plastic flows is inversely proportional to the injection time. The figures next to the points in the curve indicate the velocity of the injection screw during the injection phase. As can be seen, in order to get the viscosity low enough to fill these parts, the screw has to travel at very high velocities. To get to these very high rates of injection, higher hydraulic pressures are needed. Referring to Table 3.4, high-end injection molding machines are capable of pressures up to 20,000 psi. Thin wall moldings may require even higher pressures than this. For this, an accumulator is added to the machine, which can raise the available pressure to 35,000 psi or more. The machine can in this case have injection times as low as 0.1 seconds, allowing flow into very thin areas. As highlighted in Table 3.4, the tooling has to be a lot more robust and allow for greater venting and a very high degree of polish on the ribs for easier ejection.

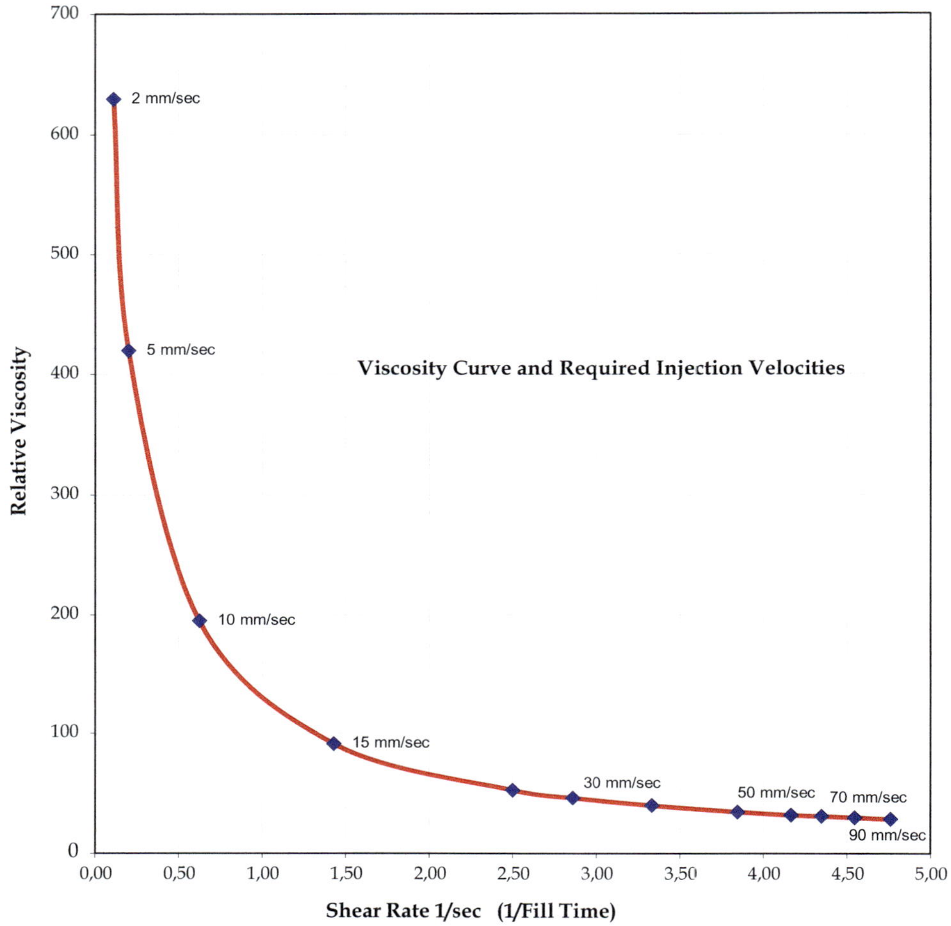

Figure 3.67 Relative viscosity vs. shear rate

Table 3.4 Special Processing and Tooling Requirements for Thin Wall Molding [9]

STANDARD VS. THIN-WALL PROCESSING			
Key Factors	Conventional	Thin-Wall	
Typical wall [in]	0.080-0.120	0.050-0.080	< 0.050
Machinery	Standard	High-end	Custom
Inject. pressure [psi]	9000-14,000	16,000-20,000	20,000-35,000
Hydraulic system	Standard	Standard	Accumulators on injection and clamp units

STANDARD VS. THIN-WALL PROCESSING			
Processing			
Fill time [s]	> 2	1-2	0.1-1
Cycle time [s]	40-60	20-40	6-20
Tooling	Standard	Better venting, heavier construction, more ejector pins, better polish	Extreme venting, very heavy construction, mold interlocks, precise surface preparation, extensive ejection features, mold costs 30-40% higher than standard

Figure 3.68 shows some typical length to width ratios for certain Covestro LLC materials.

It is best to consult the resin supplier for specific rules applicable to their resins when designing for thin walls.

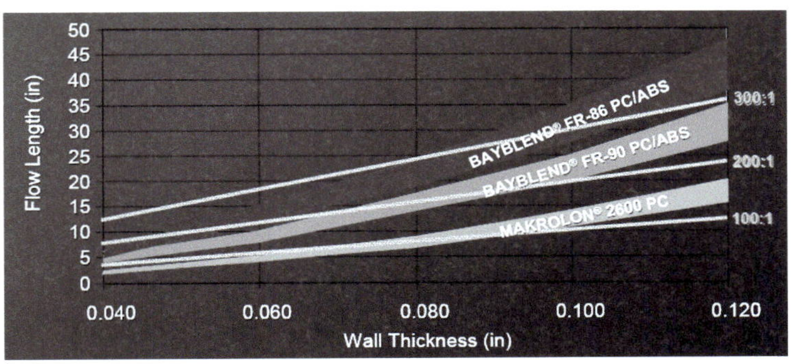

Figure 3.68 Typical length to width ratios for Covestro materials [5]

Below are some guidelines for different rules that are generally applicable to some common materials.

3.4.1.1 Wall Thickness Variation

The normal 25% variation in wall thickness and amorphous materials may result in a 1 mm wall having a section thickness of only 0.75 mm. This may be too thin to be filled.

Therefore, the wall thickness has to be kept uniform by a jog in the section that needs to be recessed. See Figure 3.69.

Figure 3.69 Jogging the wall to maintain a uniform wall thickness

3.4.1.2 Rib Thickness

The normal rule of the rib being 50 % of the wall thickness at the bottom does not generally apply to thin wall parts. This is due to the fact that the cross section at the junction of the rib in the wall thickness does not have the large differential shrinkage rate within the area as in conventional walls. This does not result in possible sink marks and associated stresses with the conventional walls. Also, with any draft a rib that is 50 % of the wall may result in a wall thickness of less than 0.5 mm at the end, which may again be too thin to fill.

3.4.2 Structural Foam Molding

In structural foam molding nitrogen gas or a blowing agent is allowed to foam the core after the material injection. This results in two solid walls on both sides of the molding sandwiching a core-like structure in between. Figure 3.70 illustrates the cross section of a structural foam part with a wall about 6 mm thick.

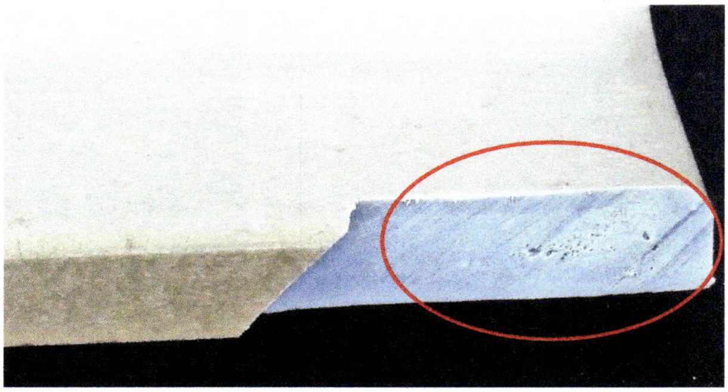

Figure 3.70 Cross section of structural foam part

Considerably thicker walls than in conventional molding can be molded with this process. Since the highest stresses in bending are those in the outer fibers, structural foam offers a greater stiffness to weight ratio than conventionally molded parts. Structural foam suffers from visual defects on the surfaces because of the foaming action and therefore needs some kind of finishing operation, such as painting, to make it visually acceptable.

Here again the wall thickness variation and rib thickness ratio conventions can be less rigorous than in conventional molding.

3.4.3 Gas Assisted Molding

In this process an inert gas, such as nitrogen, is introduced into a molded part right after injection. The gas displaces the molten core and pushes it to the extreme end of the part. This process is used where extremely thick sections are needed. As in structural foam, wall thickness variation in the rib thickness rules do not apply other than the minimum cross section required for the gas to penetrate.

Figure 3.71 shows a representative gas assisted molding part. The gate is at the very top. The gas inlet is slightly below it to the right. The gas channel formed is clearly shown from the inlet to the end of the part. Notice also that the gas does not penetrate the flat thin area at the bottom.

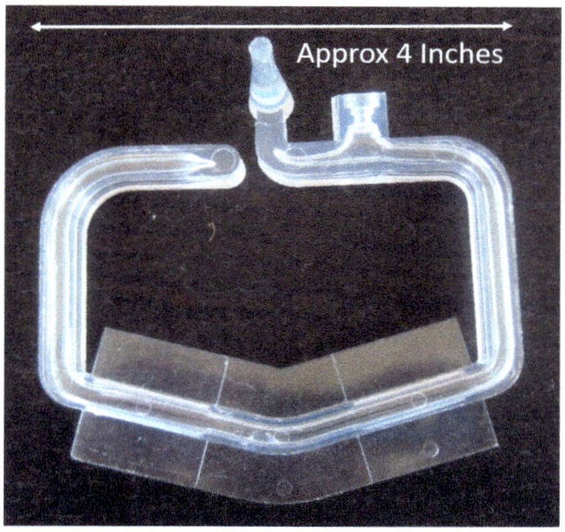

Figure 3.71 Gas assisted molding part

3.5 References

[1] Associated Press, April 1, 2014

[2] CNN Money

[3] *http://www.boeing.com*

[4] *Tim Hepher; Alwyn Scott, Reuters*, April 18, 2013

[5] Images courtesy of Covestro LLC

[6] Source: Jordan Rotheiser; *Joining of Plastics, Handbook for Designers and Engineers*, Hanser Publishers, Munich (2009)

[7] Courtesy of John Bozzelli

[8] Image courtesy of Joseph P. McFadden Sr.; *http://McFaddenCAE.com*

[9] Adapted from article by Kurt Weiss; Plastics Technology (February 2012)

[10] Suhas Kulkarni; *Robust Process Development and Scientific Molding*, 2nd ed., Hanser Publishers, Munich (2017) p. 62

[11] Courtesy of Mold Tech – Standex Engineering Group

[12] www.dfmpro.com

4 Tooling Considerations

4.1 Introduction

Continuing the theme of a holistic approach, let us now talk about the third wheel—tooling. Designers should have a good concept of the tooling basics to facilitate synergy with the tooling personnel, in order to prevent deficiencies in the area that can affect a good design and vice versa.

We will begin with the overall family of molds and proceed to some of the key features where designers need to actively participate in the tooling process.

At the highest level, an optimized mold serves three functions:

1. It is a well-designed, well-engineered, cost effective piece of machinery that needs to last a million or more cycles with minimal maintenance and be manufactured in a timely manner
2. It provides the most efficient conduit for the plastic to flow from the machine nozzle to the very end of the part and allows the proper evacuation of the air and gases from the mold cavity to the atmosphere
3. It is an advanced thermodynamic system that needs to optimize the thermal management from the hot plastic to the ejected part, delivering best possible physical properties to the part and reducing the cycle time to the minimum possible

4.2 Injection Molding Tool Family

Broadly speaking, the tool family can be divided into:

1. Two plate tools
2. Three plate tools

They can be further classified as tools with cold runners and hot runners or a combination of the two.

4.2.1 Two Plate Molds

"Plates" refers to the two blocks of steel that between them contain the cavities, cores, and other individual features that the plastic flows into to result in the finished part. The cavity creates most of the outside of the part, and the core the inside. In conventional molds, the cavity is fastened to the stationary side of the molding machine and connected to the plastic barrel through mechanisms described later in the chapter. This is also called the "A" plate.

The core or "B" plate is attached to the moving side of the mold and contains the ejector and *most* of the other mechanisms that create undercuts in the plastic parts.

Two plate molds have only one parting line. This is where the mold opens and the part is ejected from. Figure 4.1 shows the details of a two plate mold and the parting line.

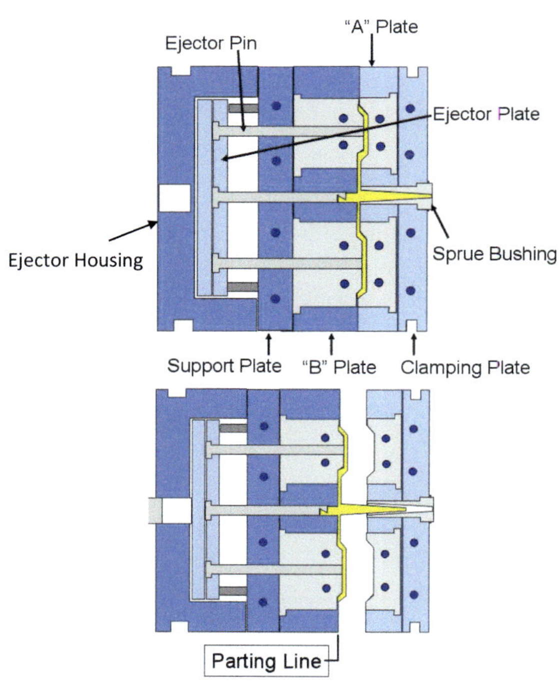

Figure 4.1 Details of a two plate mold [1]

After the mold opens at the parting line, the ejector system moves forward to release the part from the core; eventually, the part falls freely as shown in Figure 4.2 a and b. Notice the Z-shaped machined segment in the middle ejector pin. This is an undercut that makes sure that the sprue and the runner are pulled out from the cavity side along with the part. This device is called a Z puller.

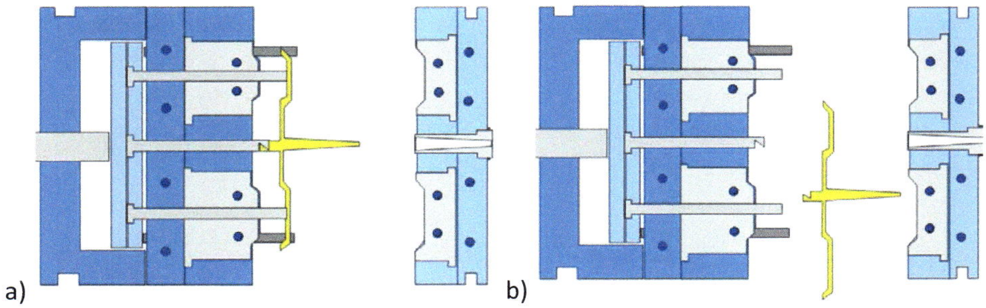

Figure 4.2 a) and b): Ejection from a two plate mold [1]

Two plate molds have the simplest construction and are generally used when the parts are gated from the side or from a central gate directly connected to the sprue.

Figure 4.3 shows a typical two plate molded part in a family mold. The main sprue is in the middle at a right angle to the runner. The runners take the plastic from the sprue to the gates. On a side note, the mold is unsuccessfully trying to balance four very different parts, as discussed later in this chapter and in Chapter 5.

Figure 4.3 A typical two plate mold part, runner, and gating arrangement

4.2.2 Three Plate Molds

Three plate molds have an extra plate called the floating plate. With the addition of the floating plate, two parting lines are created. Figure 4.4 shows the details of the three plate mold. The primary runner is contained between the floating and the A plate. The part in the A plate is fed by a secondary runner that ends up as a pinpoint gate leading to the part. The main advantage of using a three plate mold is being able to create multiple cavities with their own individual gates and/or having the flexibility to have multiple gates on one part. Three plate molds are now generally being replaced by hot runner molds that can achieve the same result with a significantly less complex mechanism and practically no runner wastage, as will be explained later in this chapter.

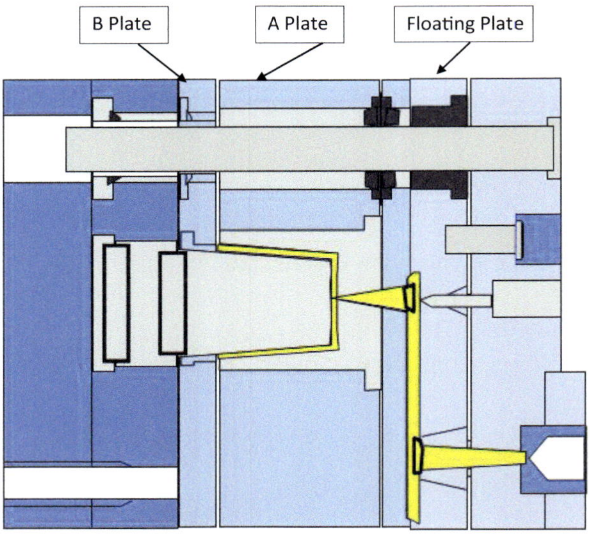

Figure 4.4 Details of a three plate mold [1]

Figure 4.5 (a, b, c) shows the sequence of ejection for a three plate mold. The mold opens at parting line 1. At a certain point the floating plate stops moving and parting line 2 opens. The ejector mechanism then pushes the part and the runner out, and the mold is ready for the next cycle. Notice that in this case the ejector for the part is a "stripper" plate. See Figure 4.40. A conventional ejector system with ejector pins can also be employed.

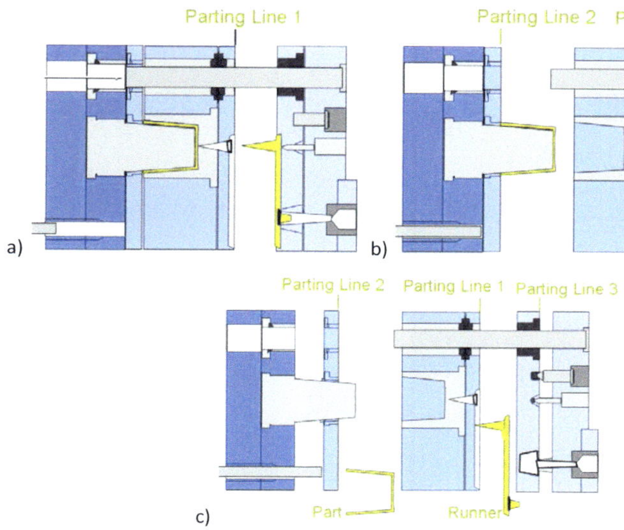

Figure 4.5 a), b), and c): Ejection sequence of a three plate mold [1]

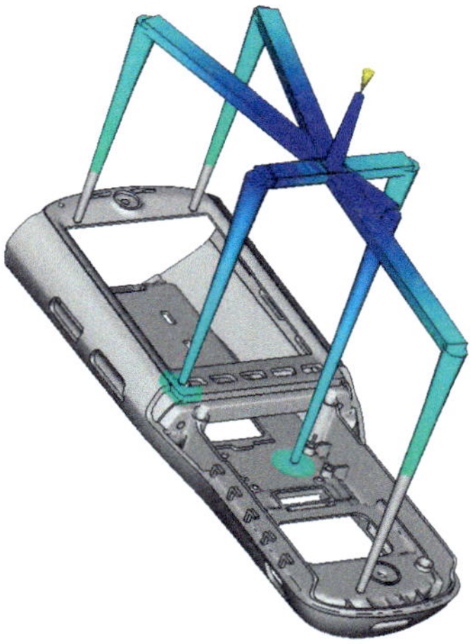

Figure 4.6 A typical part and runner from a three plate mold

4.3 Mold Elements

4.3.1 Steel Selection

There are dozens of choices for creating various components of the mold, from the mold bases to the actual core and cavity inserts. The most important factors in considering the steel selection are:

- Ease of machining
- Wear resistance
- Fatigue life (as in long pins subject to continuous back and forth movements)
- Toughness (ductility)
- Corrosion resistance
- Thermal conductivity
- Weldability
- Heat treatment requirements
- Temper
- High polish, such as for optical parts

It may seem less than obvious, but the actual cost of the material (dollar/kg) is not as important, as this is an insignificant portion of the total cost of the mold.

Table 4.1 compares the properties of the most common types of steels and aluminum from 1 to 5 (5 being the best).

Table 4.1 Comparative Properties of Common Mold Materials [8]

	Aluminum	P-20	Nak 55	D-2	S-7	A-2	O-1	H-13	420 SS
Machinability	5	3	4	1	3	3	3	3	3
Rockwell Hardness	150-170 HB	30-34	40-45	57-59	52-56	57-59	56-58	46-52	46-50
Toughness	1	2	2	4	5	4	2	2	2
Corrosion Resistance	3	2	2	4	3	4	2	4	5
Thermal Fatigue	1	1	1	1	2	2	2	5	2
Polishability	1	4	5	4	4	4	3	4	2
Heat Treatment Stability	N/A	N/A	4	4	4	4	2	3	3
Weldability	5	4	4	1	3	1	3	4	4

Effect of Mold Material on the Total Cost and Life of Molds

Although the costs of the mold materials themselves are insignificant, the machining, the added heat treatment, and subsequent grinding and polishing costs can make the overall costs very different.

As an example, the commonly used P20 materials are pre-hardened to Rockwell C 28-32 and offer a good compromise between machinability and wear resistance. H13, on the other hand, is machined "soft," hardened, and drawn down to around Rockwell C 48-56. On critical surfaces, a thin amount of steel is left that needs to be ground and polished after the heat treatment. As newer EDM (Electrical Discharge Machining) and high speed machining technologies that can perform most of the operations on the hardened steel emerge, the cost difference between the two is becoming smaller.

Below are some estimated expected lives for the various types of materials:

Aluminum (7075 T6) - 100,000 cycles

P20 or equivalent (28-32 HRC) - 300,000-500,000 cycles

H13 or equivalent (48-56 HRC) - 1,000,000 or more

Table 4.2 lists the common types of materials for various mold components, based on the country of origin and/or international standards.

Table 4.2 Materials for Mold Components [8]

COMPONENT	MATERIALS							
	AISI	Uddeholm	Assab	Werkstoff	Hasco	JIS	DIN	HARDNESS
CAVITY BLOCKS	H-13	Orvar Supreme	8407 Supreme	2344	1.2344	SKD 61	X40CrMoV5-1	Rc 46min-48
CORE BLOCKS	H-13 or S-7	Compax Supreme	Calmax	1.2358				Rc 54-56
LIFTERS	H-13	Orvar Supreme	8407	2344	1.2344	SKD 61	X40CrMoV5-1	Rc 46min-48
SLIDE FACE INSERTS	H-13	Orvar Supreme	8407	2344	1.2344	SKD 61	X40CrMoV5-1	Rc 46min-48
SLIDE CARRIERS	H-13	Orvar Supreme	8407	2344	1.2344	SKD 61	X40CrMoV5-1	Rc 46min-48
SMALL SHUT-OFF INSERTS	H-13	Orvar Supreme	8407	2344	1.2344	SKD 61	X40CrMoV5-1	Rc 46min-48
	P-20 modified	Impax Supreme	718 Supreme	2710			45CrNi6	Pre-hardened Rc 32-36
GIBS, LOCKS, STRIPPERS	A2	Rigor	XW-10	2363		SKD 12	X100CrMoV51	Rc 58-60
	D2	Sverker 21	XW-41	2379		SKD 11	X155CrVMo121	
WEAR PLATES	A2	Rigor	XW-10	2363		SKD 12	X100CrMoV51	Rc 58-60
	D2	Sverker 21	XW-41	2379		SKD 11	X155CrVMo121	
HIGH HEAT EXTRACTION MATERIALS	BeCu, Aluminum, etc.							
MOLD BASE PLATES	4130		EM33	2330		SCM430	35CrMo4	Pre-hardened Rc 28-34
	P-20+S	Holder	618	2312	1.2312		40CrMnMoS8 6	Pre-hardened Rc 31-35

AISI - American Iron and Steel Institute

4.3.2 The Feed and Vent System

The design of the feed system is very important for the efficient transfer of the plastic (softened or molten for amorphous and semicrystalline materials, respectively) from the barrel to the mold and subsequently to the part. A poorly designed system can dramatically reduce the performance and increase the cost of a part.

The following are the constituents of the feed system:

4.3.2.1 Sprue and Sprue Bushing

The sprue bushing connects the injection machine nozzle to the primary runner. The inside diameter is tapered for an easy removal of the sprue from the bushing. Figure 4.7 shows a typical sprue. The sprue "O" dimension has to be a little bit bigger than the nozzle diameter in order to get an easy removal of the sprue.

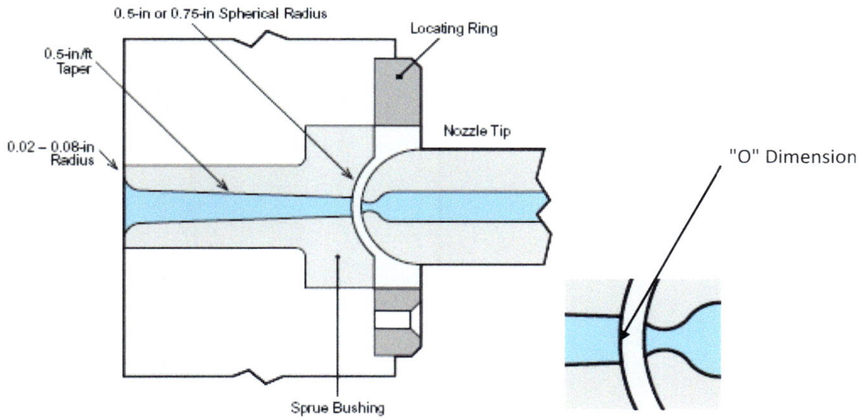

Figure 4.7 Typical sprue design [2]

4.3.2.2 Runner

The runner is the main artery from the sprue to the gate.

4.3.2.2.1 Cold Runners

Cold runner is the term used for runners that are not externally heated and are surrounded by the same temperature as the rest of the part.

The main types of cold runners are full round, trapezoidal, and half round. Figure 4.8 shows the advantages and limitations of the various runner profiles.

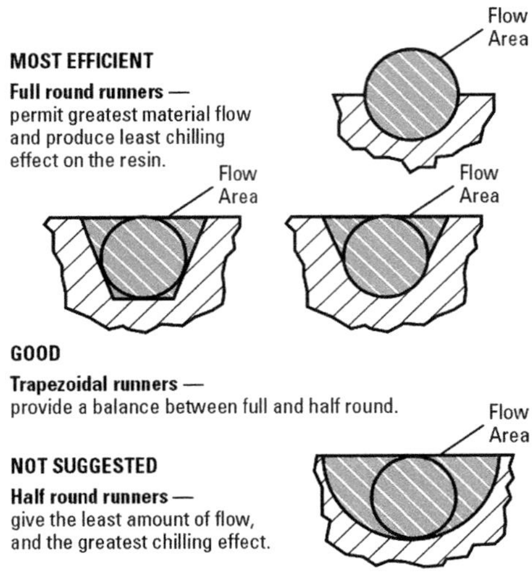

Figure 4.8 Common runner profiles [2]

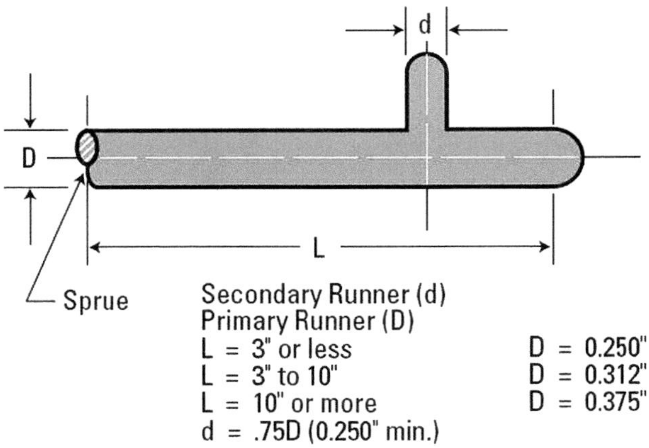

Figure 4.9 Typical dimensions for a full round runner [2]

Figure 4.10 is a typical full round runner.

Figure 4.10 Full round runner

Full round runners are recommended for two plate molds. Trapezoidal runners are more suitable for three plate molds, as they require a mechanism called "sucker pin" to make sure the sprues stay on the right side of the second parting line in order to break themselves off from the gates. The trapezoidal cross section facilitates creating a large pad to create the sucker pin area.

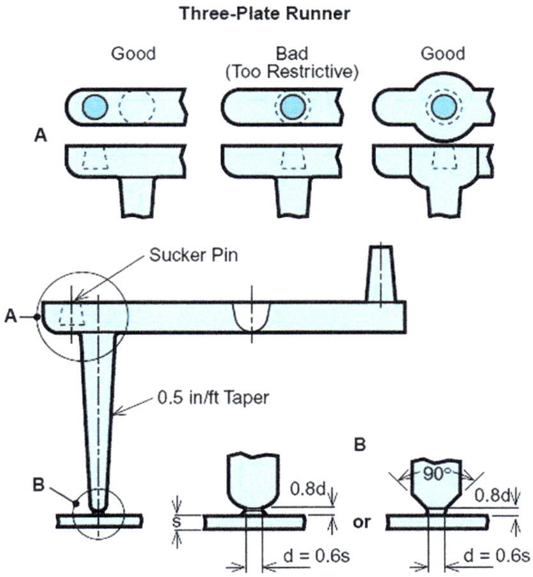

Figure 4.11 Three plate runner and gate design [2]

Figure 4.12 shows a typical three plate molded part and runner system. The mold plates have been removed for clarity.

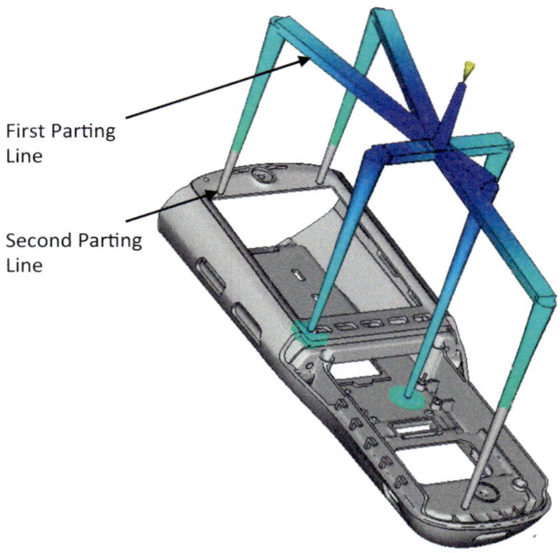

Figure 4.12 A three plate molded part and runner (mold plates removed)

The restrictions caused by poorly designed sucker pins can cause serious yield and part performance issues. See Figure 4.13 as an example. The sucker pin restriction is starving the part and causing the main runner to splash.

As a result:

- There is a lot of unnecessary scrap
- The molding machine has to use an unnecessary amount of energy to generate the higher pressures to fill the part
- There is shot to shot variation in the parts
- The parts are underpacked and the knit lines are very weak

Figure 4.13 An actual runner with poor sucker pin design

4.3.2.2.2 Multicavity Molds

Multicavity molds can save processing costs, provided the parts are the same or fairly similar in filling requirements to each other. Figure 4.14 shows some configurations of naturally balanced runners.

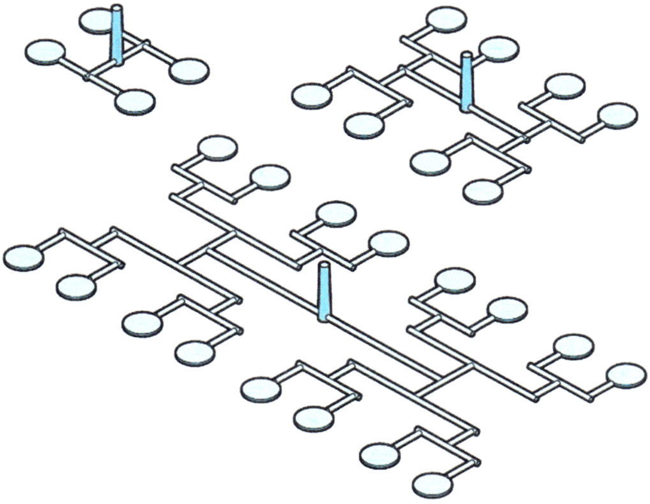

Figure 4.14 Some naturally balanced runners [2]

A mold with cavities for non-identical parts is known as a family mold. These are generally not economical, as–in order to balance the filling pattern of the different parts–none of the parts end up being optimized for processing. It is not uncommon to have one or more cavities blocked off to improve the overall yield. Figure 4.15 shows a family mold for producing test strips for mechanical and other testing of various resins. Since the parts have very different volume and pressure requirements, some of the cavities are being overfilled and are flashing while others are starving. A common mistake is to try to adjust the gate size to balance the filling. This is wrong on two important counts:

- For a part of a given thickness, there is *only one* optimum width or diameter
- Gates are very sensitive to thickness variations and the slightest error can impact the filling efficiency

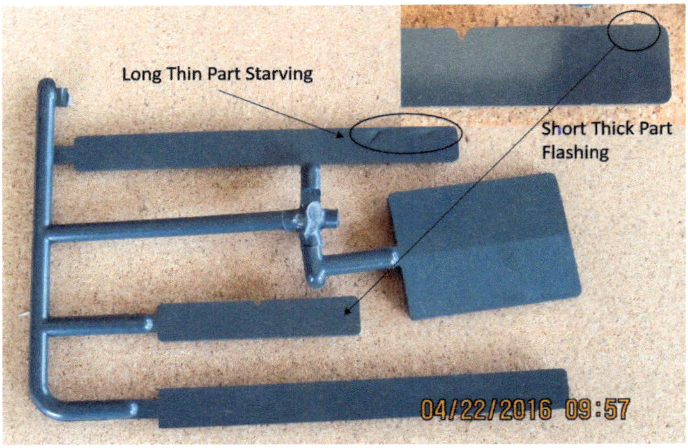

Figure 4.15 Unbalanced family mold

Multicavity Molds–Special Considerations

A word of caution here. It is very common to have the inside cavities filling more than the outside ones when there are more than four cavities. Dr. John Beaumont has done some remarkable research on why this happens. Figure 4.16 shows an actual part and runner system showing the inside cavities being filled faster than the outside ones. This situation is corrected by using a "Melt Flipper®[1]" in the mold.

1 Melt Flipper is a registered trade name owned by Beaumont Technologies, Inc.

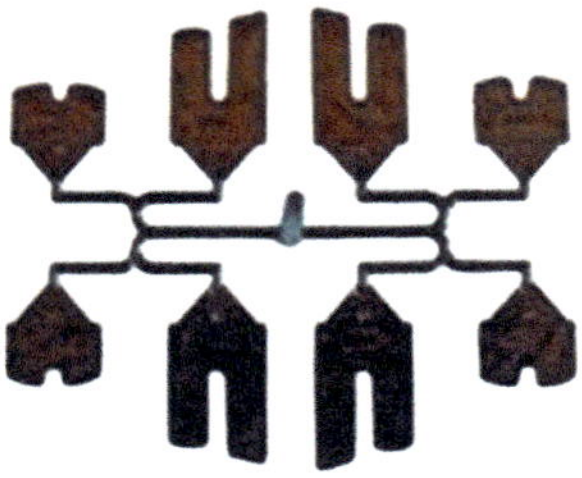

Figure 4.16 Actual parts showing inside parts being filled faster than the outside ones [7]

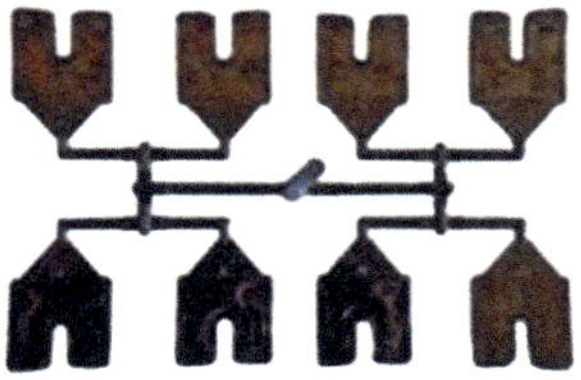

Figure 4.17 Actual parts showing balanced filling with Melt Flipper [7]

See Figure 4.18 for an explanation. High-temperature, low-viscosity material resulting from shear thinning and heating flows along the outside perimeter of the primary runner, while cooler, high-viscosity material remains inside. When the flow hits the first runner branch, the hotter, more fluid material stays close to the inside wall of the secondary runner, while more viscous material flows around it and adheres to the opposite runner wall. The result is that the inside cavities fill faster with the hotter, less viscous material, while the outer cavities get shortchanged by the slower, cooler material [7]. It should be expected that parts formed with these different melt conditions will shrink and warp differently.

The imbalanced melt conditions discussed above will develop anytime a runner branches and therefore can create problems in molds with as few as two cavities. As illustrated in Figure 4.18 a and b, after only a single branch the parts may fill at the same time, but the hotter and the cooler material will be distributed to opposite sides of each cavity. This can create residual stresses and warpage.

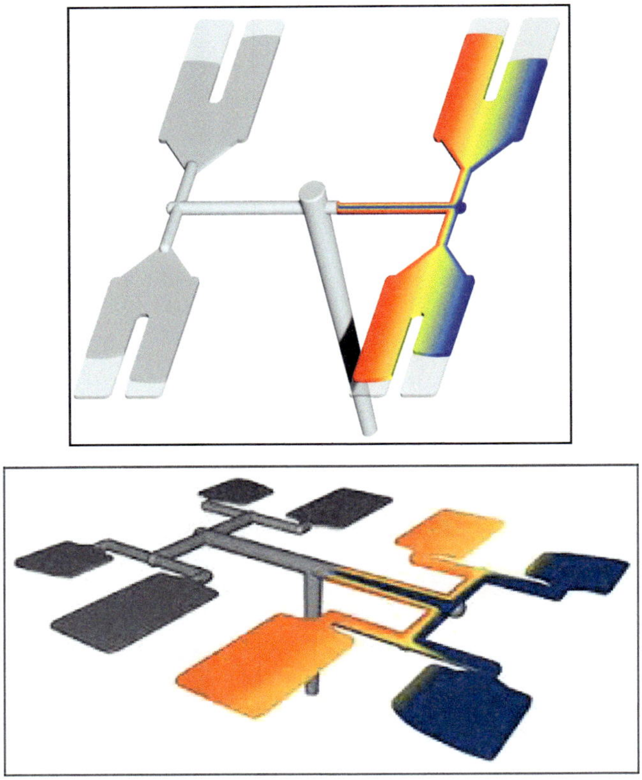

a)

b)

Figure 4.18 a) and b): Explanation of why inside cavities fill faster [7]

4.3.2.2.3 Hot Runner Systems

Hot runner systems are the ones that are locally heated by some type of external heating arrangements, most typically by high resistance wires wound around them. The runners are always in the molten or softened stage. When the first cycle starts, material fills the runner to fill itself and the part. Once the part is filled and packed, the flow of the plastic into the runner stops. In the second and subsequent cycles, the material in the runner from the previous cycle starts to fill the part and more material flows through the runner to complete the filling of the part and the runner.

The primary advantage of hot runners is that there is no material wasted as in the cold runners. The other advantage is being able to have multiple gate locations into the cavity, just as in the three plate mold, without the complexity of the three plate mold and the maintenance of the moving parts.

Some of the disadvantages of hot runners are the controls required to maintain the temperature required, possible traps, and hot spots where the material can continue to stagnate over many cycles, degrading the material. They may also be more expensive than the three plate mold with a longer lead time needed to design, fabricate, test, and deliver the system. Also, once designed, the gates cannot be increased or decreased nor their locations changed.

Figure 4.19 illustrates a typical hot runner system. A hot sprue bushing (as opposed to the conventional or cold one discussed earlier) is heated by a heater around it. The material is then pushed into the heated channels serving as runners. The material is finally delivered to four "drops" that lead to the surface of the part or gates.

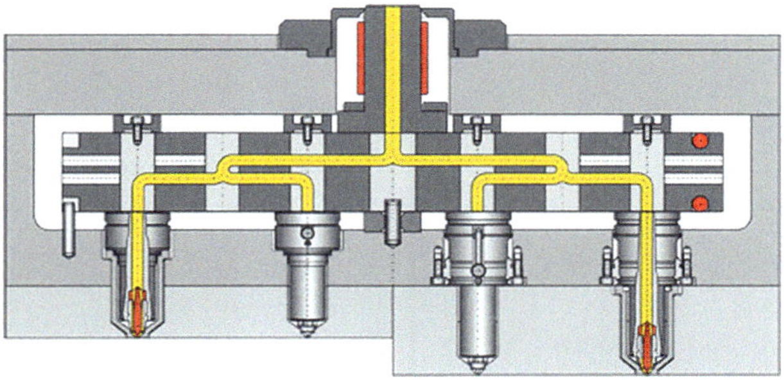

Figure 4.19 A typical hot runner system [4]

Sometimes, a cold runner system may require more material than the part itself. In this case the material savings can be enormous.

Figure 4.20 shows one such part where the savings were extremely impressive. In this mold, the change of the cold runner to a hot runner resulted in much better processing, resulting from a more balanced filling along with lower plastic pressures. The runner weighed 30 grams and used a PC from SABIC costing about $8 per kilogram. Each runner was about 25 cents in material cost. The thickness of the runner required an additional time of 20 seconds to make sure it could be ejected without getting tangled in the mold. The additional cooling time cost 33 cents at a machine rate of $60/h. Thus the extra cost per part was 58 cents. For 1 million parts, the total cost was $580,000. Note that this does not include a yield improvement of about 5% resulting from the change, as the shot to shot variation from under and overpacking was eliminated.

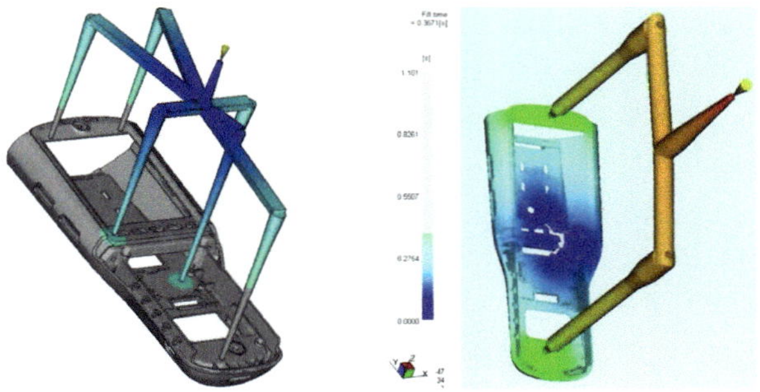

Figure 4.20 Cold runner system changed to hot runner

4.3.2.2.4 Cold Slug Wells

Between cycles, the tip of the material from the nozzle gets cold and can cause aesthetic or physical defects in the part if allowed to enter the part. Therefore, there is a section in the runner system that traps the cold material and prevents it from going into the part. This is called the cold slug well and is shown in Figure 4.21. Cold plastic material flows to these sections and gets trapped instead of going into the part.

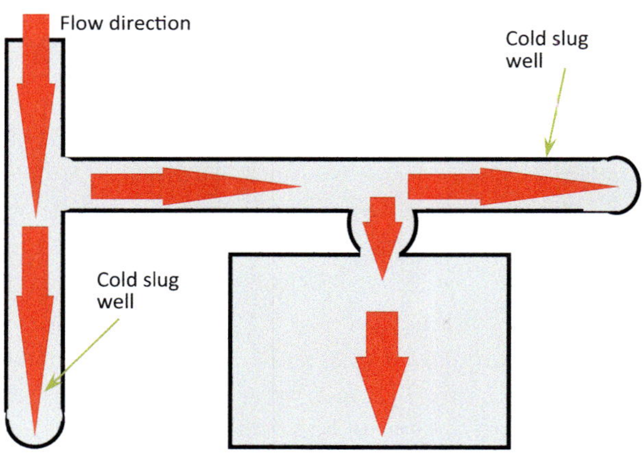

Figure 4.21 Cold slug well [2]

4.3.2.3 Gates

Gates are the entry point to the part and, undoubtedly, one of the most important parts of the feed system. They have the following requirements:

- Have large enough cross section(s) and optimized location(s) to allow flow to the end of the part with a reasonable plastic pressure
- Have a small enough cross section that will "freeze" in a reasonable time to prevent the backflow of plastic

- Be located in the thickest area of the part to facilitate thick to thin flow
- Be in the right number for a complete filling of the part
- Being one of the weakest areas in the part, not to be located in an area that will see higher stresses
- Not to have a visually objectionable vestige
- Be self de-gating or allow easy and clean gate removal
- Not to allow "jetting"—a visual snake-like appearance that is both a cosmetic and a physical defect. See Figure 4.22 (more discussion to follow in the processing section, Chapter 5). For this reason, the gate should be perpendicular to the runner as shown in Figure 4.23.

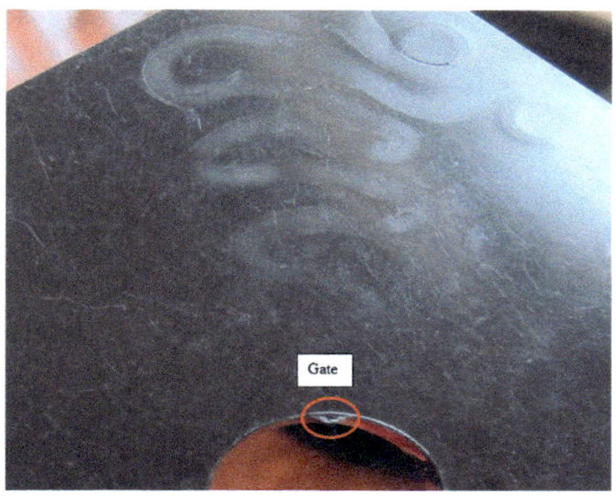

Figure 4.22 Parts showing jetting

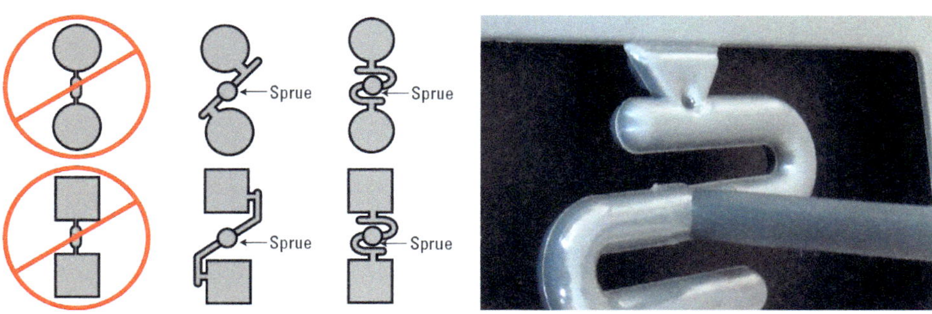

Figure 4.23 Proper gate design to avoid jetting

4.3.2.3.1 Types of Gates

Figure 4.24 shows most of the common types of gates and they are fairly self-explanatory. Tunnel gates have one extra advantage in that they are self de-gating. They can be gated directly into the part or onto an ejector pin shortened or modified for the purpose.

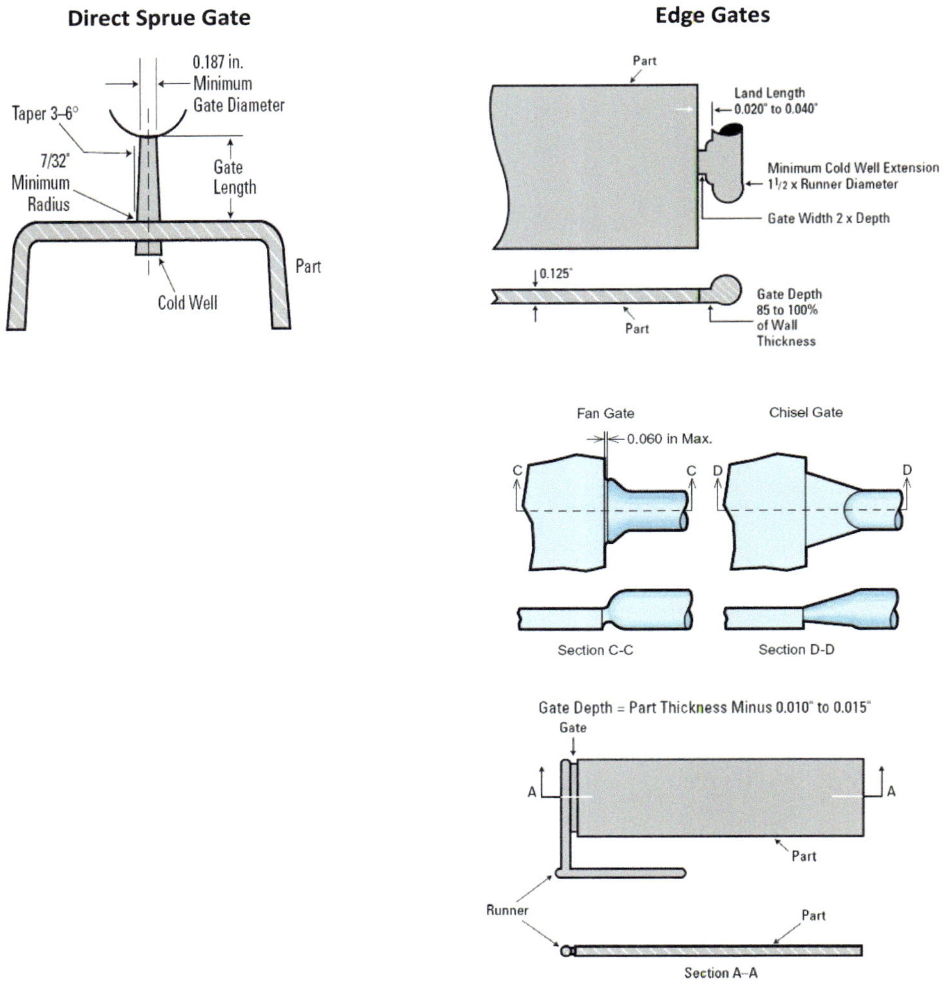

Figure 4.24

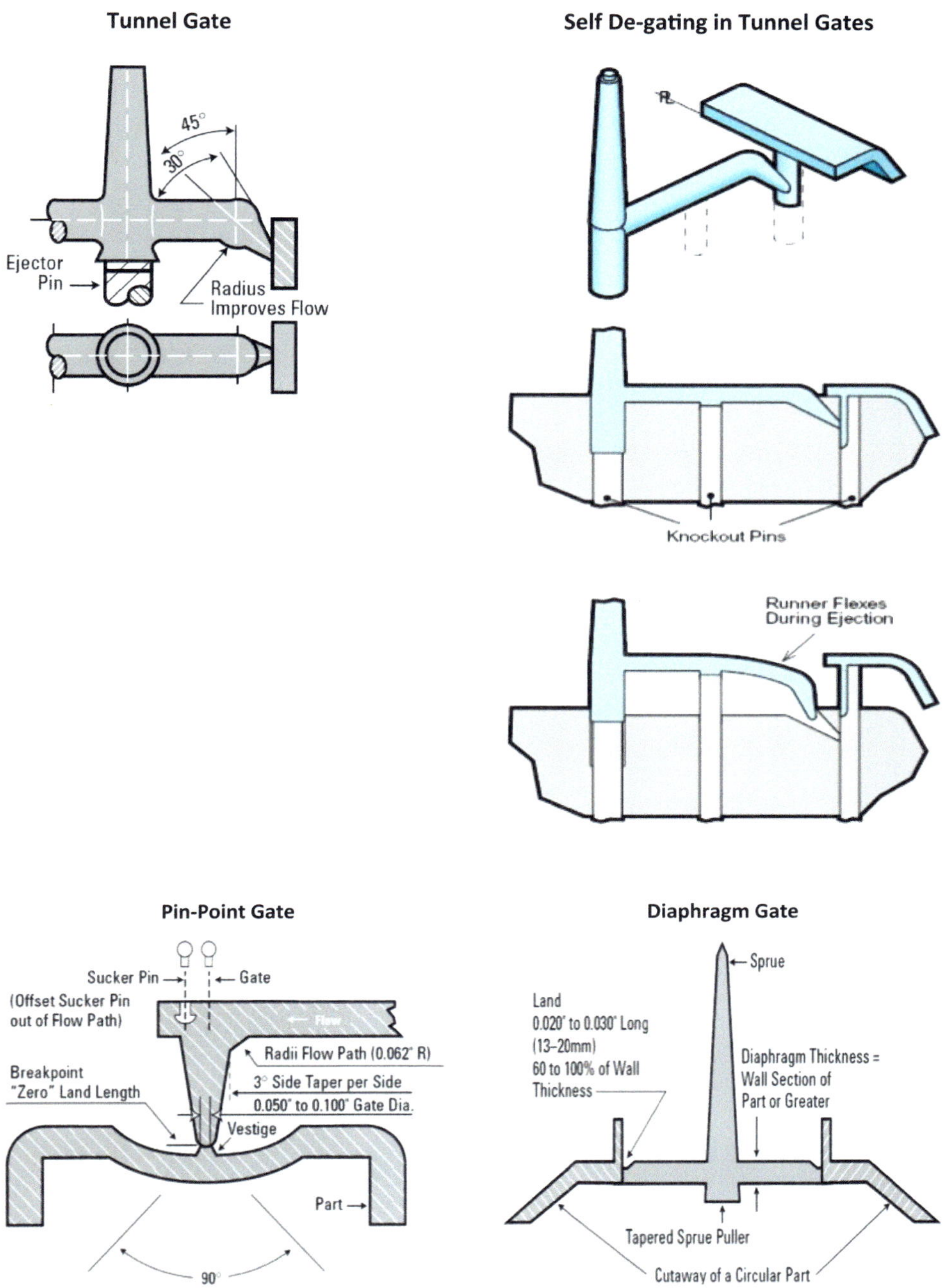

Figure 4.24 Various types of "cold" gates [2]

4.3.2.3.2 Hot Sprue Gates and Valve Gates

Hot sprues are used when the distance from the nozzle to the gate is such that the material will get too cold to flow, as in the case of reverse injection molds. They have a heater around the flow channel to maintain the processing temperature of the material right up to the gate. They are also used where the gate vestige has to be kept to a minimum for aesthetic reasons.

A variation of this gate is the valve gate, for a positive stop of the flow from the gate for aesthetic or processing reasons.

The gate diameter should be 0.156-0.181 inches for unfilled semicrystalline materials and 0.187-0.250 inches for filled ones. Additionally, for amorphous materials, there should be a 7 to 9 degree taper and zero land as shown in Figure 4.25.

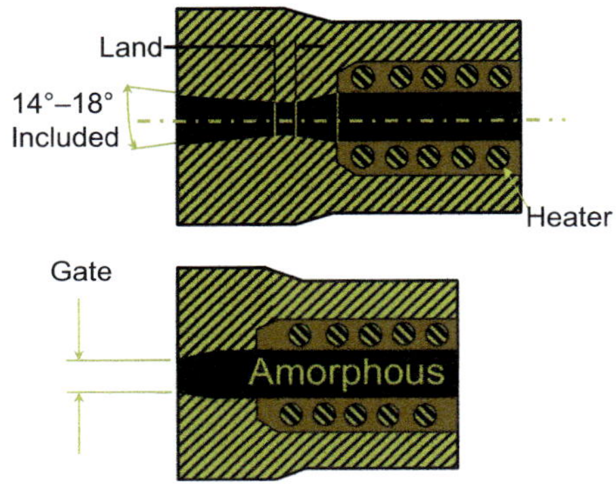

Figure 4.25 Hot tips and gates [2]

4.3.2.4 Vents

Complementing the feed system in optimum filling is venting. As hot material enters the mold, it needs to replace the air that is present in the cavities. Additionally, the material itself may generate gases that need to be expelled. Going back to Chapter 3 Section 3.2.6.3.2, if the air and gases are not expelled, they will be compressed and cause issues, not the least of which are burn marks.

Figure 4.27 shows a boss that was not properly vented at the end of the fill. It trapped gases within the boss that caused it to fail prematurely.

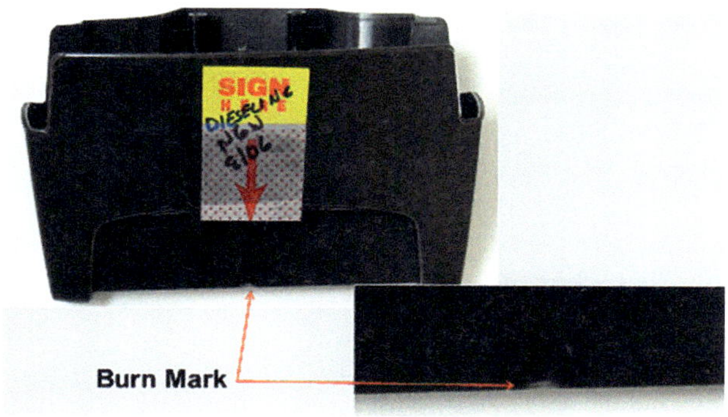

Figure 4.26 Burn marks left by trapped gases

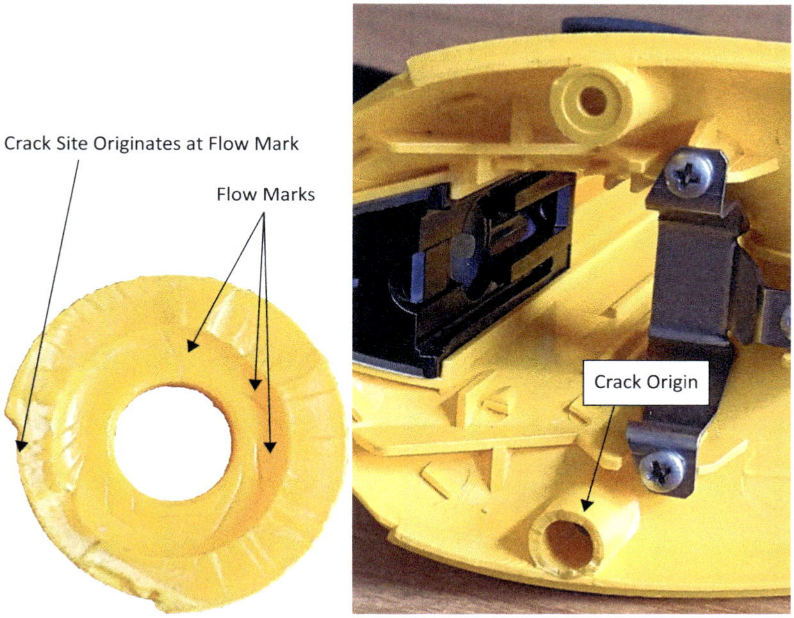

Figure 4.27 Boss broken off due to trapped gases

The vent depth needs to be a compromise between preventing material flashing through the opening and having a large enough opening to allow the rapid evacuation of the accumulated gases. In general, semicrystalline materials flow more easily than their amorphous counterparts. For this reason, the vents for the former are shallower than the ones for the latter. Table 4.3 shows the recommended vent depths for some of the most common materials.

Table 4.3 Recommended Vent Depths for Various Materials

Polymer	Depth of Vent [in]
ABS, SAN, HIPS	0.002
Acetal	0.0005
Acrylic	0.002
Cellulose/Acetate/Butyrate	0.001
Nylon 6/6	0.0005
Polycarbonate	0.0015
Polyethylene	0.001
Polypropylene	0.001
Polyethylene Oxide	0.001
Polyphenylene Sulfide	0.0005
Polysulfone	0.001
Crystal Polystyrene	0.001
PVC (Rigid)	0.001
PVC (Flexible)	0.001

4.3.2.4.1 Number, Position, and Cross Sections of Vents

The material pushes the trapped gases to the end of the flow and therefore it is important that there be plentiful vents at the end of the flow. As in the case of the draft, the more, the better.

In order to achieve a rapid evacuation of gases, it is also important that there be the least resistance to the flow of air. Therefore, after a controlled "land" the depth of the vent needs to be increased. Figure 4.28 illustrates the spacing, location, and cross section of the vents.

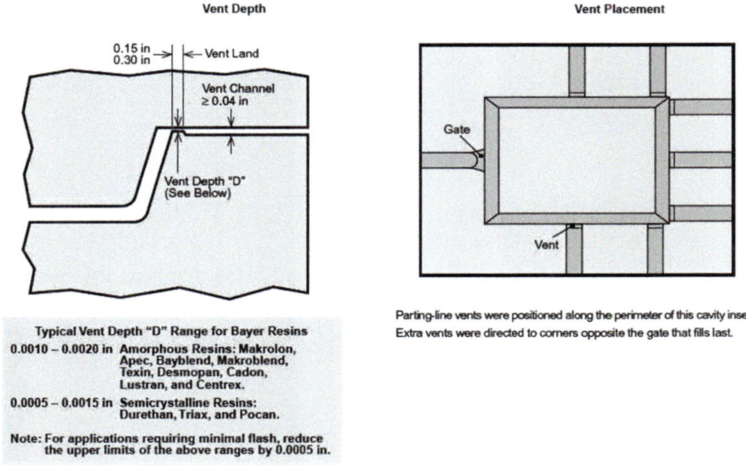

Figure 4.28 Vent layout and cross section [2]

4.3.2.4.2 Vented Ejector Pins

In order to provide venting on the middle of the parts, vented ejector pins can be used on both sides of the part. Since the pins move back and forth on the core side, they are self-cleaning. On the cavity side the vents are stationary and will get clogged over time, and have to be periodically removed and cleaned. Vented pins located directly in the area of knit lines will aid them to be visually and structurally more sound. See discussion on 3.2.6.2, Chapter 3, for more details. Figure 4.29 shows a vented ejector pin.

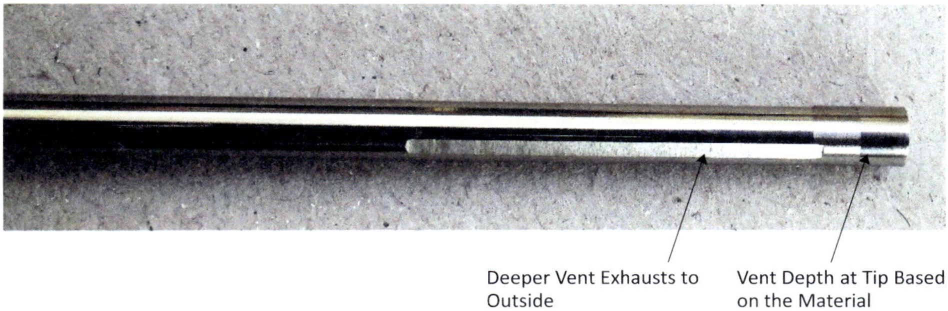

Figure 4.29 Vented ejector pin

4.3.2.4.3 Porous Materials

Porous materials are available that can be used literally as "sieves" on the surface of the mold. Inserts made from these materials can be used on functional parts for rapid venting. In cosmetic areas, the surface created by the porous steel may not match the surrounding area. Also, witness lines will be left around the insert.

Typical porous metal parts used in mold vent applications range in diameters from 0.125 inches to 0.500 inches and have a thickness range of 0.062 inches to 0.125 inches or greater. These inserts will get clogged over time and may need to be removed and cleaned at regular intervals. Figure 4.30 shows a highly magnified (100×) view of the cross sections of two of these steels from Mott Corporation[2].

[2] An informational brochure from one of the manufactures of porous metals, Mott Corporation, is available at *https://mottcorp.com/sites/default/files/Porous%20Metal%20for%20Mold%20Vents.pdf*

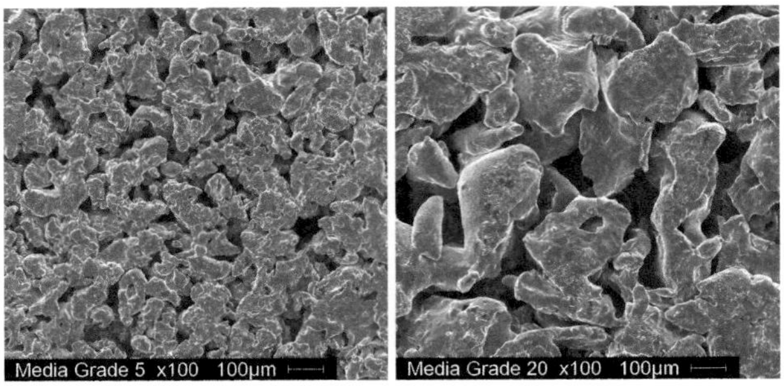

Figure 4.30 SEM images of porous metal 316LSS (left) and 3166LSS (right) [5]

4.3.3 Creating Undercuts

More often than not, plastic parts have male or female features that are in a direction not parallel to the direction of draw (direction in which the mold opens). The term for these is undercuts. In order to be ejected, the steel forming the feature needs to move out before the ejectors move forward. Listed below are some of the most common ways to achieve this.

4.3.3.1 By Pass Shut Off

When the design allows for it, this is the simplest way to create an undercut. Figure 4.31 shows how undercuts can be molded without the need for a moving piece of steel. The design needs a hole or slot in the part to achieve this, as shown. Also, the angle between the mating faces of steel needs to be a minimum of 5° to 7° in order to allow free movement and for prevention of galling between them.

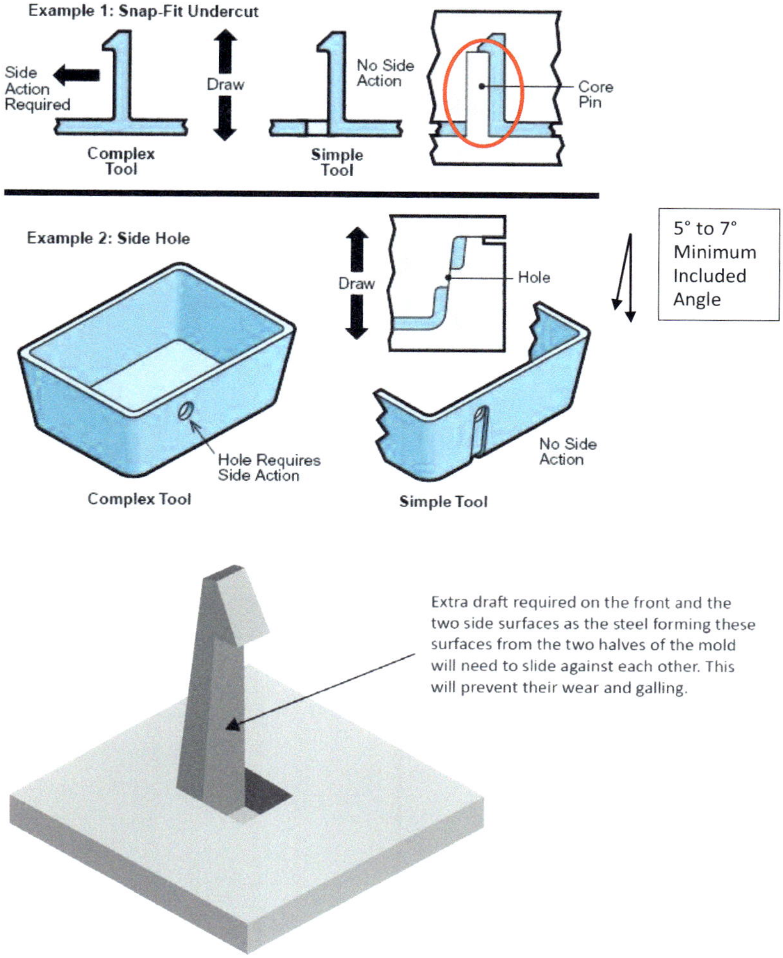

Figure 4.31 Alternate ways to create undercuts if the design allows; top image courtesy of Covestro LLC [2]

4.3.3.2 Angle Pin or Finger Cam

A moving block (slide) is made to move laterally as the mold opens and closes. This is achieved by having the block containing the feature ride on a cam pin, which is at a certain angle. The angle is calculated based on the amount of lateral movement needed. The angle pin is fixed to one half of the mold. Thus when the molds open, the pin activates the slide perpendicular to the direction of the mold opening. It is important to note that the pin only facilitates the lateral movement. A firm heel block is required to provide the positive stop to maintain the part dimensions by not allowing any movement when the mold is closed.

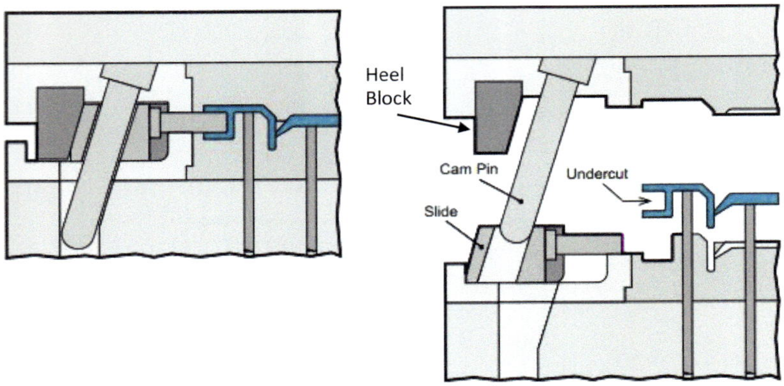

Figure 4.32 Angle or finger cam [2]

4.3.3.3 Hydraulic Cylinders or Solenoids

Hydraulic cylinders or solenoids are alternates to the finger cam, when the angle exceeds 30°. They are also used to form undercuts on the cavity side, as the steel members creating the undercut have to be completely retracted before the mold opens.

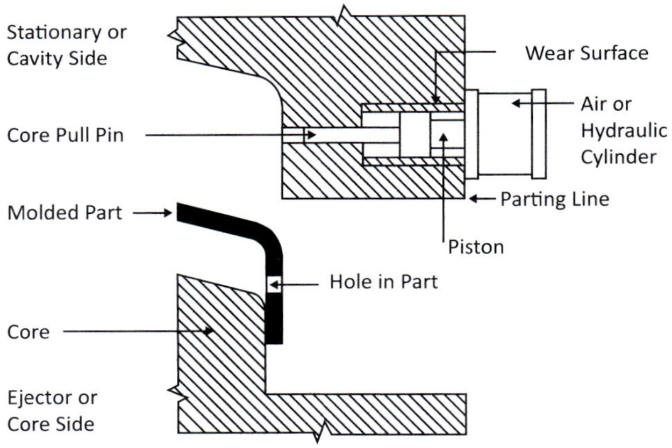

Figure 4.33 Hydraulic cam in action

4.3.3.4 Floating "A" Plate

Shallow undercuts on the cavity side can also be formed by a mechanism called a floating "A" or cavity plate. In this case, the cavity separates from the mounting plate and moves with the core for a short distance, allowing the lateral movement of steel members forming the feature on the cavity side to release the undercuts.

4.3.3.5 Collapsible Core

Collapsible cores are used on threaded parts where the undercuts are internal to the part, such as bottle caps.

The mechanism consists of four or more spring loaded "petals" that, when expanded by a cam pushing from the inside, form a complete circle that the threads or other similar features have been machined on. Before the mold opens, the cam moves back and the petals collapse, releasing the parts from the undercuts. See Figure 4.34. The bottom images show the open (left) and closed (right) petals.

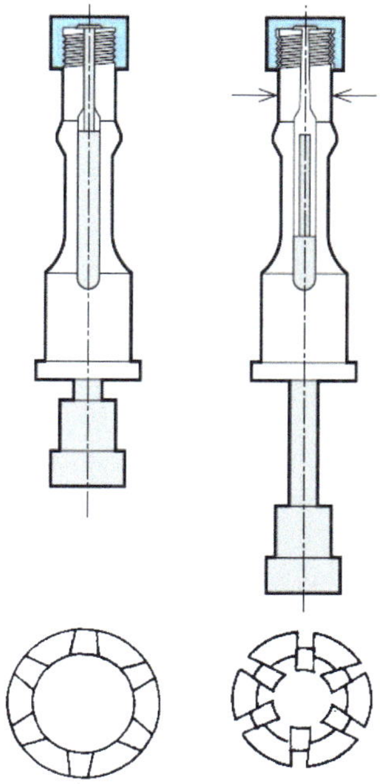

Figure 4.34 Collapsible cores; top images courtesy of Covestro LLC [2]

4.3.3.6 Lifters

Lifters are one of the most common ways to create internal cuts on features such as snaps.

A lifter rides in an angled bushing and moves laterally when the mold opens or closes. The undercut feature is machined on the outside face of the lifter. See Figure 4.35 for details. Notice that since the lifter face rubs against the plastic surface along the length of its travel, no features or graphics can be designed in this length and width (X) of the lifter.

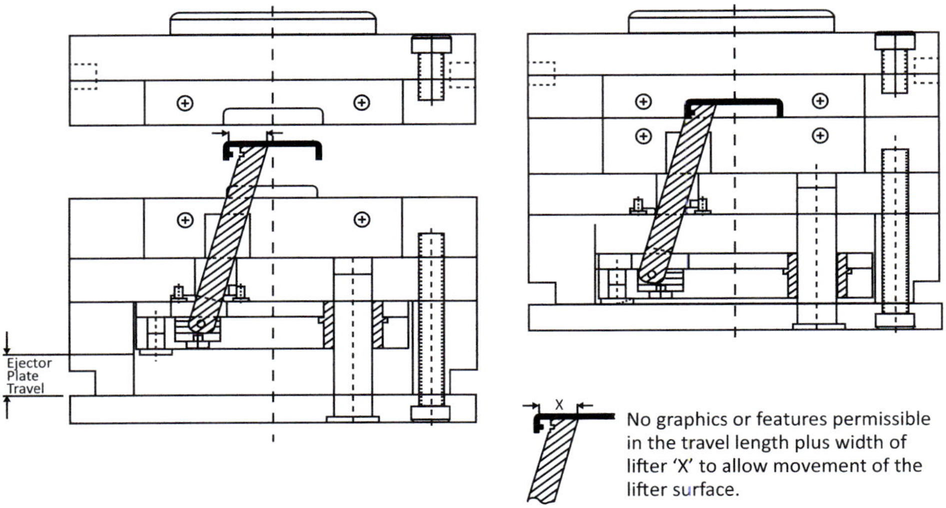

Figure 4.35 Lifter and lifter mechanism

4.3.3.7 Loose Piece Insert

For shorter runs and to expedite a prototype mold, loose piece inserts can be used to create both inside and outside features. These pieces are manually put into the mold before it closes and they are ejected along with the part. There are generally two or more of these, depending on the cycle time. The second piece is inserted into the mold while the insert is being separated on the work bench by the operator after a cycle, to maintain a uniform molding cycle. A robot or an operator is needed to place and remove the insert during the molding cycle.

4.3.3.8 General Notes on Side Action

While side actions and parts with undercuts are ubiquitous, they should preferably be avoided where the design allows it, due to the following reasons:

- They add to the cost of the mold, both initially and for ongoing maintenance for wear and tear
- They add to the complexity of the mold
- Whenever two pieces of moving steel form a feature on a part (as opposed to the feature being formed by solid stationary pieces of steel), variation gets added and the tolerances have to be loosened

4.3.4 The Ejector System

The plastic shrinks anywhere from about 5 to 20 times or more than the steel it is formed on, resulting in a very tight grip onto the steel. Thus, it needs an ejector system to push it out uniformly and with an adequate area of support from the ejectors. The force has to be balanced so that the entire part comes out uniformly without being distorted. It needs to have an adequate area of contact to make sure the ejectors do not penetrate the part.

The following are the constituents of the ejector system:

4.3.4.1 Ejector Pins and Blades

Ejector pins and blades push the part off the core as the ejector plate moves forward. Ejector pins generally have a round cross section and are used where there is an adequate area in the plastic to overlap it. Ejector blades are used in areas where there is only a limited area of engagement on the part, such as the top of ribs. Both ejector pins and blades leave witness marks on the part. Blade ejectors are more difficult to fit and maintain. Figure 4.36 shows some typical pins and blades. They basically consist of a cylindrical pin with a shoulder that is used to assemble them onto the ejector plate, as shown in Figure 4.1 and Figure 4.37. The pin has a larger diameter in the back for strength and stiffness, and is machined down to the hole diameter in the front that will match the hole in the core block. The diameter is reduced over a length that will be adequate for the stroke of the ejector. In the case of the blade, the pin is ground to the cross section of the slot in the core block.

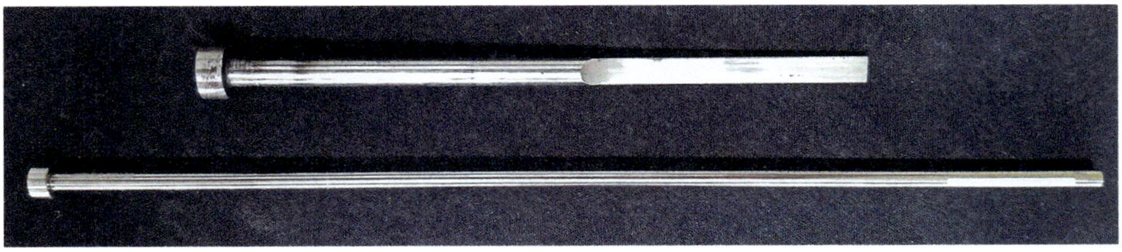

Figure 4.36 Ejector pin on the bottom and blade ejector on the top

Figure 4.37 shows a complete assembly, including the part before and after the ejection. The pin is to the left and the blade is to the right. Notice that the blade on the right is completely covered by the plastic wall whereas the pin on the left is only partially covered. The arrangement on the left may lead to the pin penetrating and distorting the part.

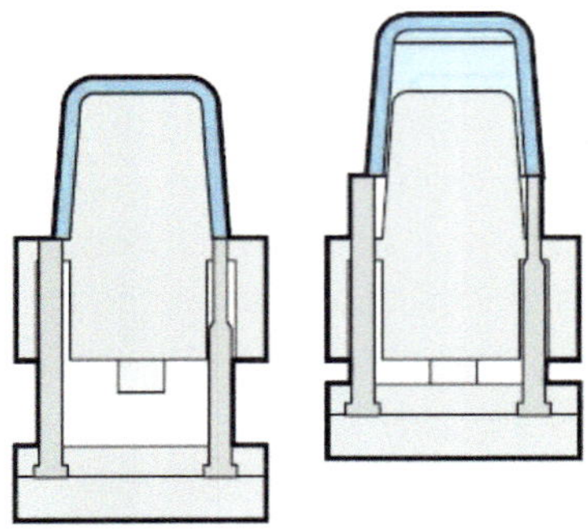

Figure 4.37 Ejector pin on the left and blade on the right [2]

4.3.4.2 Ejector Sleeves

Ejector sleeves are used both to create the hole in and eject a hollow boss. Looking at Figure 4.38, the pin floats inside a sleeve. The shoulder of the sleeve is sandwiched in the ejector plate and the shoulder on the pin is fixed to the ejector housing (see Figure 4.1). The pin is the diameter of the hole in the boss and creates it. The sleeve is the diameter of (or slightly smaller or larger than) the boss.

Upon ejection, the moving sleeve pushes the boss out. Figure 4.39 shows an actual example.

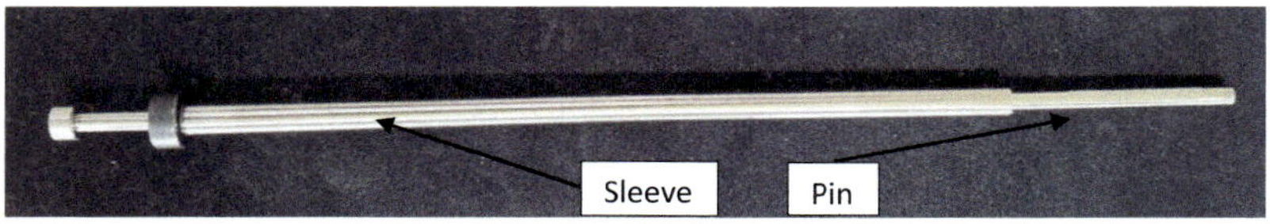

Figure 4.38 Sleeve ejector

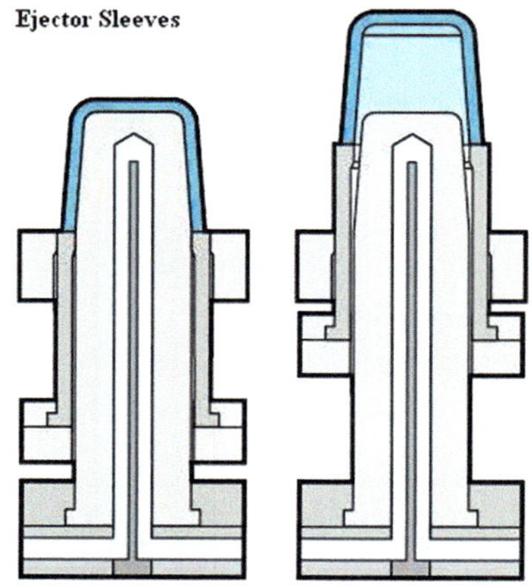

Figure 4.39 Sleeve ejector in action [2]

4.3.4.3 Stripper Plate Ejection

In large hollow parts such as a cup, a solid ring around the core is a separate piece connected to the ejector plate and moves with it. Thus, the whole periphery of the part is pushed out by this ring, called the stripper plate. See Figure 4.40. Stripper plates do not leave a witness mark.

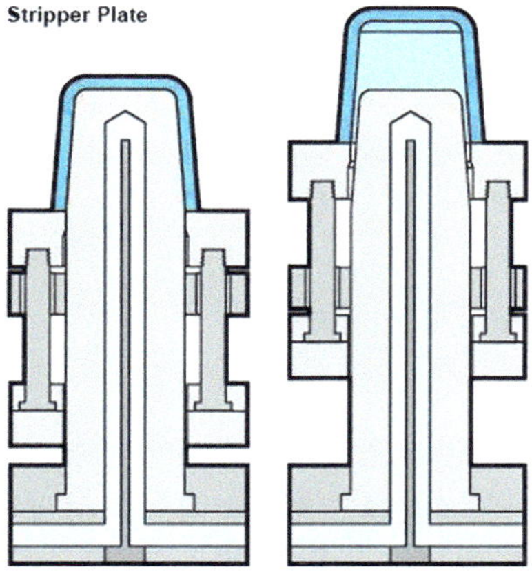

Figure 4.40 Stripper plate before and after ejection [2]

4.3.4.4 Air Poppet Valve

Again, in a large hollow part such as a cup, high pressure air can break the vacuum and push the part out. This is accomplished by a valve that fits into the core and seals against it during injection and cooling. When ejection is required, compressed air is pushed through this valve and causes the part to be pushed out. See Figure 4.41.

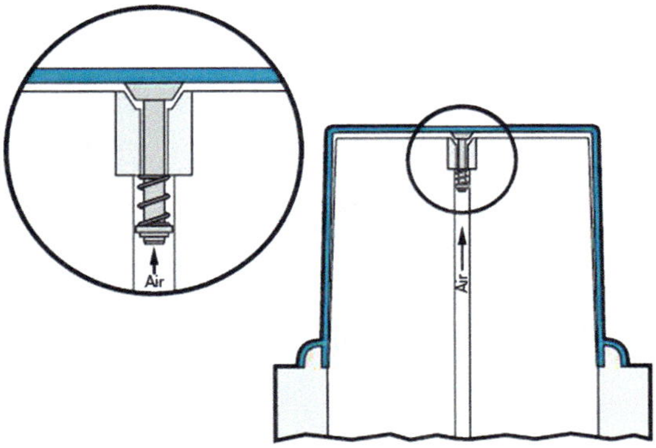

Figure 4.41 Air poppet valve [2]

4.3.4.5 Factors Influencing Ease of Ejection

The most important factors are:

- Draft on the part
- Part geometry that allows for relatively large diameter ejector pins and blades uniformly around the part
- Mold polish (polishing in the direction of draw helps ejection)
- Plating on the mold, such as PTFE impregnated nickel or chrome
- Material characteristics
  - Semicrystalline materials like nylon are inherently more lubricious than amorphous materials such as PC
  - Non-lubricious materials can have additives such as PTFE and silicon to help the release
- Modulus of elasticity: while a higher modulus of elasticity keeps the material from distorting and being able to be ejected sooner, it also suffers from not being flexible and not being forgiving to intentional and unintentional undercuts
- Cooling time and ejection temperature: as the part gets colder, it becomes stiffer and aids the ejection without distortion; however, this also means that it has shrunk more on the core and therefore is more difficult to eject

4.3.4.6 Guided Ejection

A guided ejection system consists of a series of robust pillars attached to the ejector housing going through high wear bushings in the ejector plate assembly. This makes the ejectors always move parallel to the B plate, in turn ensuring a long life for the system and preventing uneven wear of the holes in the core block that can otherwise become elongated and cause flash.

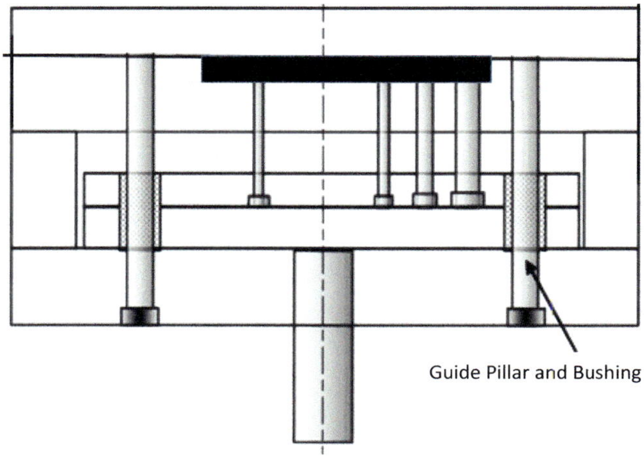

Figure 4.42 Guided ejector system [3]

4.3.5 Special Case: Reverse Mold

Normally, the gate is attached to the cavity half or at the outside edge of the part. Every once in a while there is no place to have a gate vestige on the outside surface, due to cosmetic or functional reasons. In these cases, the gate needs to be on the core side that is attached to the stationary half of the machine. Since there is no movement on the core side in this case, the ejection has to be achieved with hydraulic cams or solenoids. Look at Figure 4.43. The part is gated directly on the inside. Notice the two large hydraulic cylinders on the sides that operate the ejection system. Also notice the long sprue. Since it is so long, it needs a heater band around it to maintain the proper injection temperature for the plastic. This is called a heated sprue.

The bottom view in Figure 4.43 shows a large reverse molded cover. It was reverse molded because it had a central gate and the gate vestige could not be on the front of the cover, due to obvious cosmetic reasons.

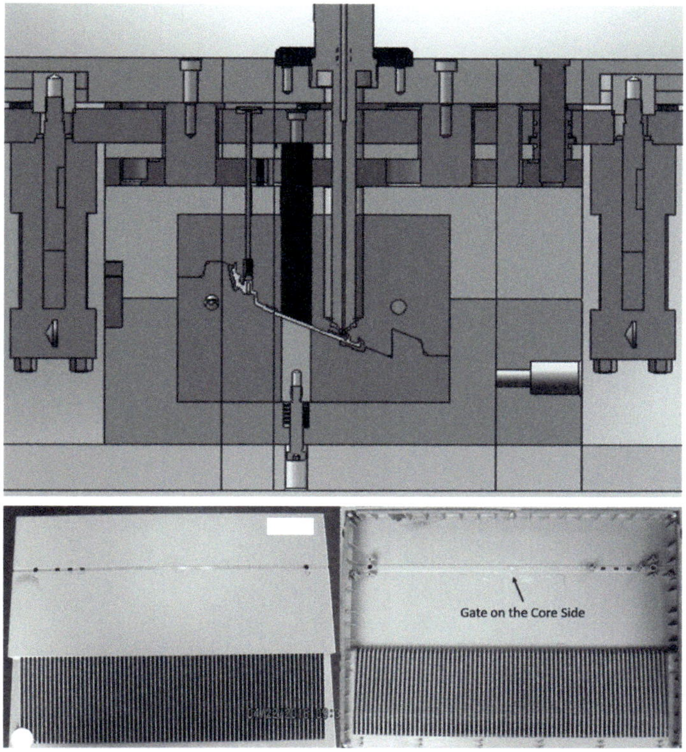

Figure 4.43 A reverse injection mold and part

4.3.6 Thermal Management

I remember one of my first classes in Internal Combustion Engines and the words of my professor, "The purpose of the coolant in the engine is not to cool the engine, but to optimize the temperatures outside the cylinders for the best thermal efficiency of the system as a whole."

Although not nearly as complicated as the internal combustion engine, the mold is a complex thermodynamic system requiring well engineered thermal management. Modifying my professor's words, "The purpose of the cooling lines in the mold is not to cool the mold, but to optimize the temperatures in the mold for the best plastic properties and thermal efficiency of the system as a whole." As I mentioned in the beginning of this chapter, it needs to manage the heat transfer from the hot plastic to the ejected part in order to maximize the required physical properties and reduce the cycle time to the minimum possible.

Indeed, one can run chilled water through the cooling system for very fast molding cycles with very little warpage and sink marks that are pushed under the surface. However, the performance of the parts from such a process will take a nose dive. Some high temperature materials such as Rynite PET resin from Dupont require that the molds be run as high as 120 °C, and they need oil and not water to heat the molds.

Unfortunately, this area is the worst managed and least understood in the tooling and molding industry. More often than not, a lot of "seat of the pants" designs are used and, consequentially, not optimized.

Once the "molten" plastic material reaches the core and cavity inserts that form the part, it starts to cool down immediately. Only sufficiently cooled parts can be ejected from the tool without distortion or damage from the ejector mechanism. The rate at which the parts cool down is very important to define their mechanical properties as well as aesthetics. Rapidly cooled parts may have a high level of residual stresses and warpage. Uniform cooling improves the part quality by reducing residual stresses and maintaining dimensional accuracy. The required mold temperature is even more critical to semicrystalline materials. The right combination of cooling time and mold temperature allows the crystalline areas to go back to the orderly state within a short time after ejection, and it facilitates dimensional stability and the required physical properties.

In terms of time, the cooling portion may account for more than half of the total cycle time. See Figure 4.44. An efficient cooling circuit reduces the cycle time and increases the overall productivity.

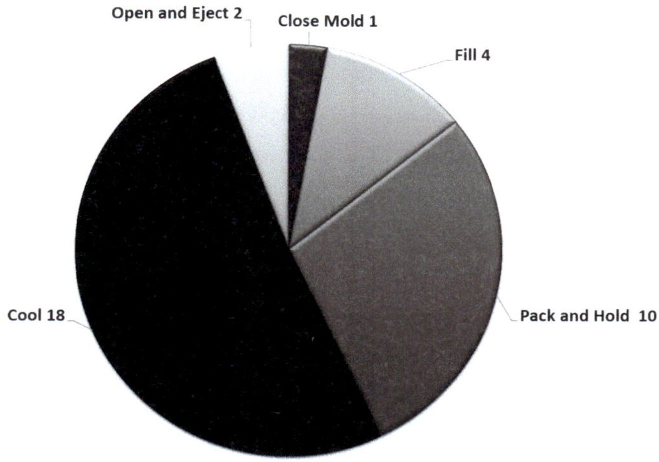

Figure 4.44 Typical injection molding cycle

An ideal cooling system has cooling lines or mechanisms as close to the mold wall as the structural requirements of the wall demand and should uniformly cover both sides of the part, including ribs and bosses. In hard to reach areas, beryllium copper inserts with their higher thermal conductivity can be used to facilitate the heat transfer. The temperature difference between the inside and the outside of the part should be minimal and preferably not exceed 10 °C upon ejection for parts with tight tolerance requirements.

Some typical cooling layouts are shown in Figure 4.45. Multiple series may offer the best option for the control of temperatures in specific areas, for a complex part with challenging mold temperature requirements.

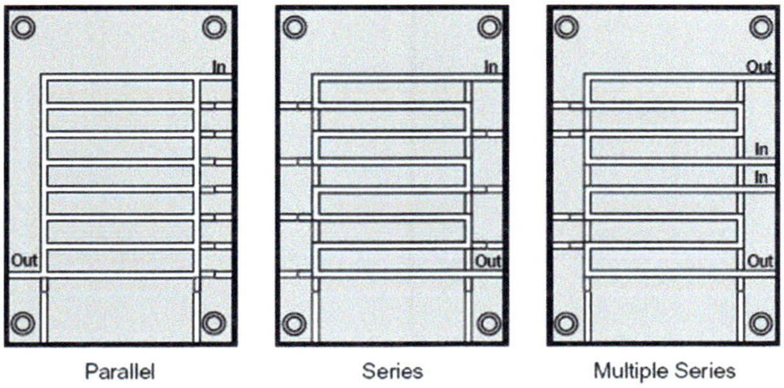

Figure 4.45 Typical cooling line layouts [2]

Figure 4.46 illustrates the typical diameters and placement of cooling lines with respect to the mold surface.

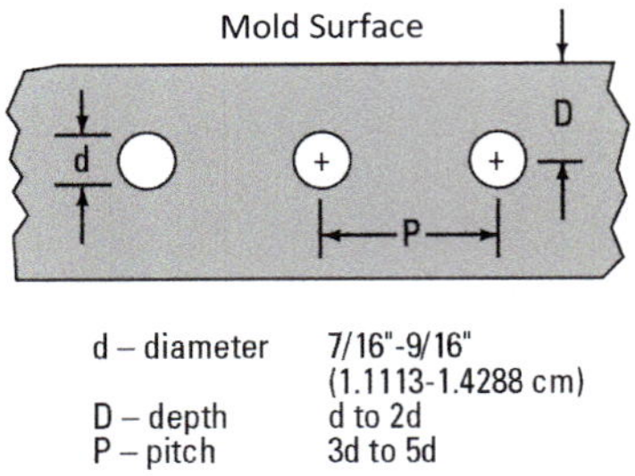

Figure 4.46 Typical cooling line area and placement [2]

Cooling lines offer the least complex way of cooling (or heating) a mold. Where cooling lines are not possible, other mechanisms to cool the mold exist. Some of the most common ones are provided in the following pages.

4.3.6.1 Cooling Insert and Baffles

This insert has spiral cooling circuits around it with sealing O-rings to prevent water leaks. It is used in deep hollow parts as shown in Figure 4.47.

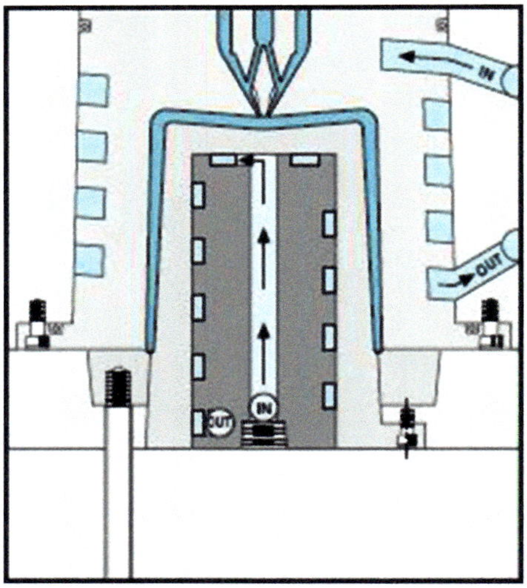

Figure 4.47 Cooling inserts [2]

4.3.6.2 Bubblers

As the name indicates, water is made to bubble up a narrow channel and back in areas with narrower cores than above. See Figure 4.48.

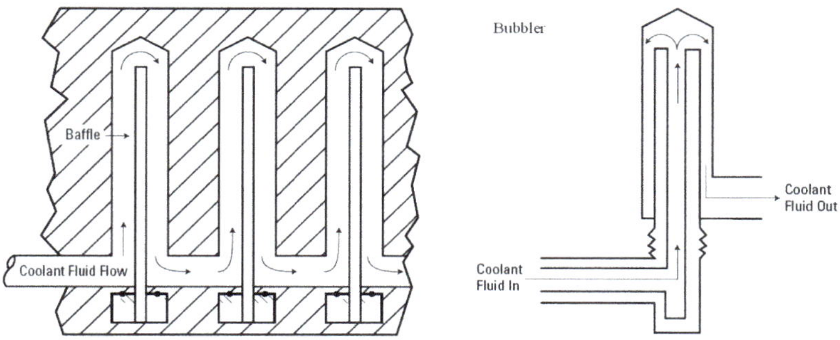

Figure 4.48 Bubblers [2]

4.3.6.3 Thermal Pin

If the cross-sectional area of the core is too small for cooling inserts, baffles, or bubblers, thermal pins or a heat pipe can be employed as shown in Figure 4.49.

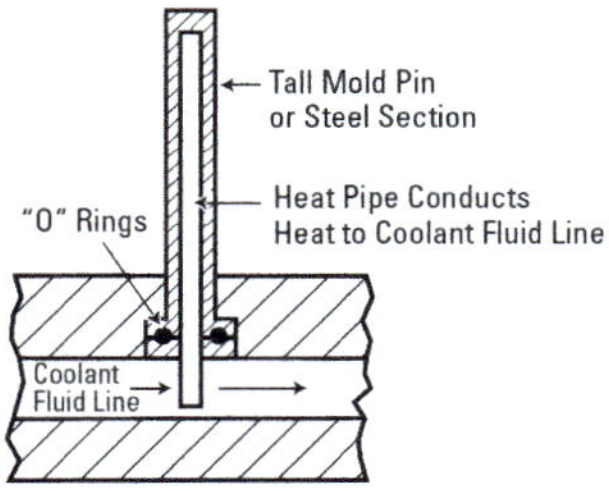

Figure 4.49 Placement of thermal pin or heat pipe in a long thin core [2]

Looking at Figure 4.50, a thermal pin consists of a sealed container (usually made of copper), a wicking structure, and a working fluid. The wicking material provides a capillary effect that allows the fluid to travel along the wall of the container. The working fluid, under its own pressure, enters the pores of the capillary material, wetting all internal surfaces. Applying heat at any point along the surface of the thermal pin causes the liquid at that point to boil and enter a vapor state. When that happens, the liquid picks up the latent heat of vaporization. The gas, which then has a high pressure, moves inside the sealed container to a colder location, where it condenses. Thus, the gas gives up the latent heat of vaporization and moves heat from the input to the output end of the heat pipe. The condensed fluid travels back along the wick and repeats the process. Thermal pins have an effective thermal conductivity significantly greater than that of copper, while weighing far less [6].

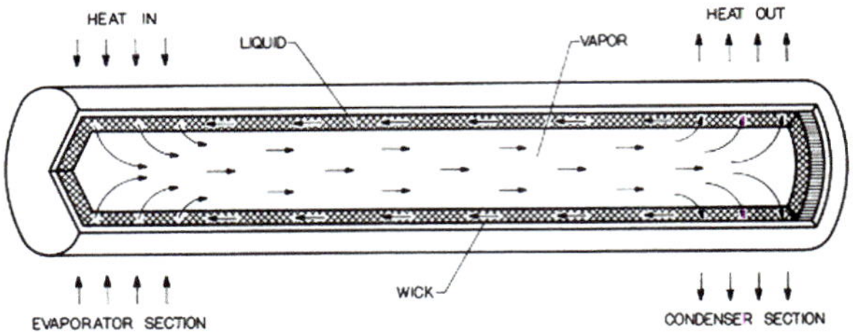

Figure 4.50 Cut section of thermal pin [6]

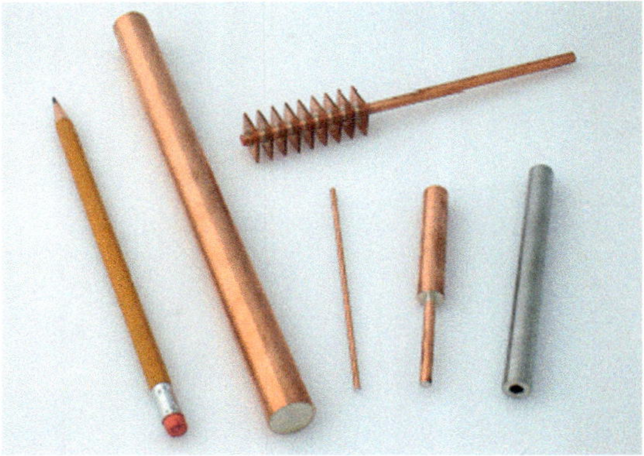

Figure 4.51 Actual thermal pins juxtaposed with a pencil to show size [6]

4.3.6.4 Coolant Flow Requirements

Almost all molds, other than those for very high temperature resins, use water because of its high heat absorptivity and ready availability.

For effective cooling, a certain flow rate has to be maintained for a maximum heat transfer from the steel to the water. Those with a mechanical engineering background should be familiar with laminar and turbulent flow. As a refresher, see Figure 4.52. The top is laminar flow and the bottom, turbulent. Water itself is a relatively poor conductor of heat. In laminar flow just the small volume of water touching the outer wall of the hole is in contact with the source and therefore it removes very little heat. The inside layers flow through without absorbing any heat.

The Reynolds number (Re) distinguishes laminar flow from turbulent and depends on the flow rate for a certain orifice size and temperature. The following are the numbers for water at 20 °C:

Re < 2300: Laminar

2300 < Re < 4000: Transient

4000 < Re: Turbulent

Figure 4.53 shows the coefficient of heat transfer vs. the Reynolds number. It is obvious that the flow (in liters or gallons per minute) should be such that a Reynolds number > 4000 is achieved, for an efficient cooling.

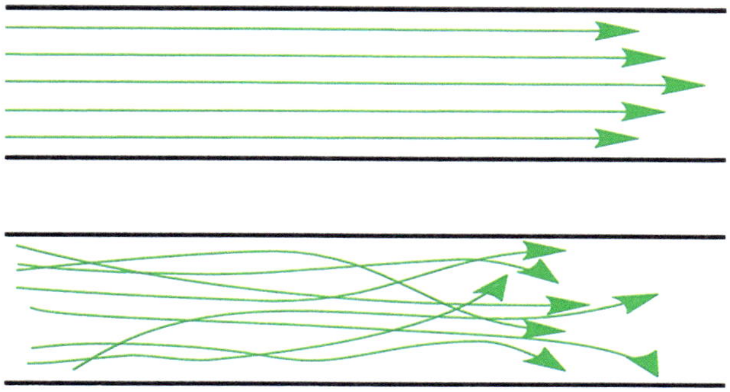

Figure 4.52 Visualization of flow; top: laminar, bottom: turbulent

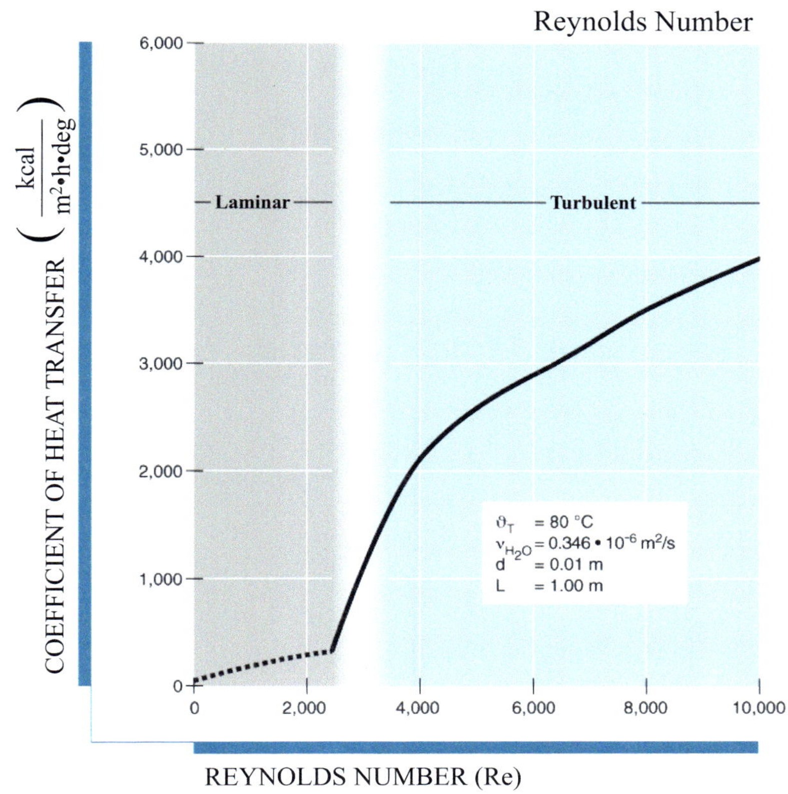

Figure 4.53 Coefficient of heat transfer vs. Reynolds number [2]

4.4 Steel Safe Condition

As will be discussed in Chapter 9, the total tolerance stack-up of a molded part comes from the tool dimensions, actual shrinkage of the plastic, and the final processing conditions.

When the tolerance requirements are extremely critical, as in the mating surface of a snap or external reveals, extra steel can be left on the steel creating the feature. This is called a "steel safe" condition. Once the actual dimensions are obtained from an optimized molding set up with the right resin, this feature can be machined to the nominal dimension required, minimizing the tolerance build up.

See Figure 4.54. Additional steel is left at the tip of the snap during the initial mold building. After the first parts are molded, the snap width can be adjusted to the exact requirement by machining off the steel, which gets replaced by plastic.

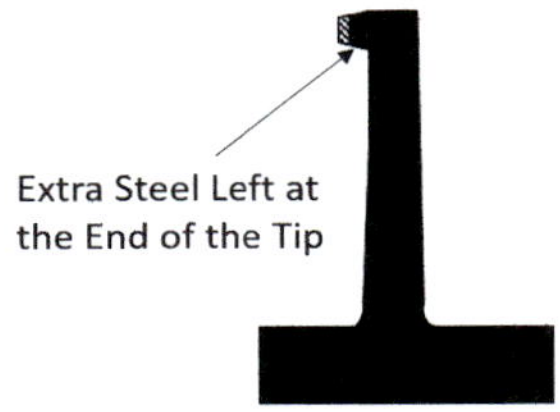

Figure 4.54 Steel safe left on snap feature and adjusted after molding trials

4.5 Reviewing Mold Designs

Although conventionally not the designer's job, a basic understanding of the documentation for mold designs is useful to facilitate the holistic approach.

4.5.1 Information Needed

The tool drawings should consist of:

- Title block
- Core and cavity plan views
- Sections passing through all important features
- Gates-enlarged view with dimensions
- Ejector housing and ejector locations

- Cooling lines layouts
- Side action-detailed section views showing clearances of metal with part
- Sections passing through shut off and parting line

4.5.2 Title Block

The title block should contain the following information:

- Material type and grade name
- Tool steel for core and cavity
- Hardness and heat treatment details
- Shrink rate for the resin–for semicrystalline or filled materials, both the flow and cross flow direction
- Press size required (tonnage, platen size, and opening)
- Part number and revision level
- Number of cavities

4.5.3 Core Plan View

The core plan view should contain the following information:

- Ejector pin locations
- Cooling lines
- Part number and date stamp location
- Gate(s) plan view and location
- Critical steel tolerances
- Number of cavities

4.5.4 Cavity Plan View

The cavity plan view should contain the following information:

- Cooling lines
- Vents
- Gate plan view and location
- Critical steel tolerances
- Enlarged gate(s) view consisting of (see Figure 4.55)
  - Gate location

- Gate dimensions
- Gate land length
- Runner dimensions

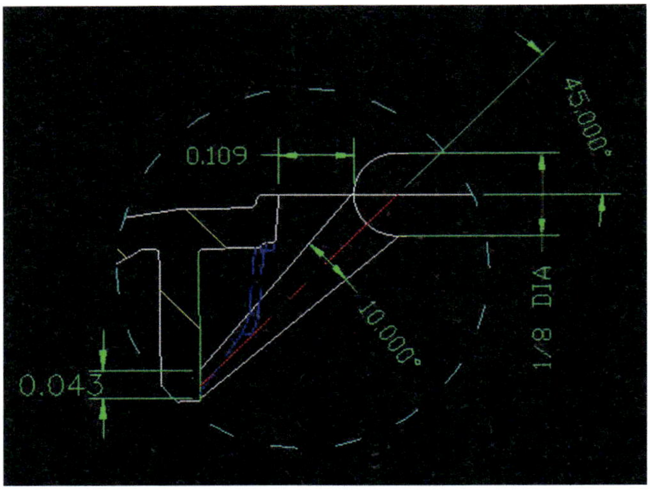

Figure 4.55 Typical gate details 4:1 view

Vents details (see Figure 4.56) should have

- Vent layout
- Locations
- Depth and width
- Land details
- Relief details

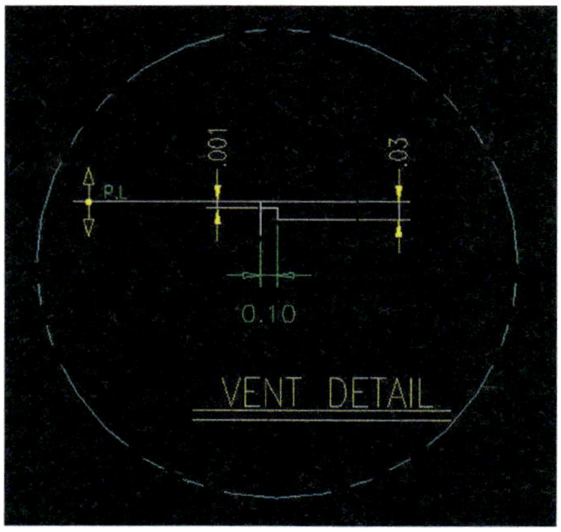

Figure 4.56 Vent details

4.5.5 3D Rendering of Part

The 3D rendering of the part should show:

- Parting lines
- Gate location(s)

4.5.6 Mold Design Checklist

Ideally, there should be an agreed upon checklist that the toolmaker fills and verifies against mutually agreed upon items, consisting of the following as a minimum (see Chapter 7, Section 7.3.1, for an example):

- Plastic, resin, supplier, grade, color
- Shrinkage
- Cavity and core steel combination and their respective hardness
- Runner balancing verified using flow simulation without obstruction from items such as sprue pullers
- Adequate venting of cavity and core
- Adequate cooling verified by flow simulation
- Side actions have a 5 HRC difference between mating surfaces

A detailed checklist can be downloaded from *www.vikpedia.org*.

4.6 References

[1] Image courtesy of SABIC

[2] Image courtesy of Covestro LLC

[3] Adapted from *http://moldright.com*

[4] *http://www.ewikon.com/en/products/hotrunner-manifolds.html*

[5] Courtesy of Mott Corporation, Farmington, CT, U.S.A.

[6] Noren Products Inc.; *http://www.norenproducts.com/sites/default/files/pdf/tm/Noren_Thermal_Pin_Catalog.pdf*

[7] Courtesy of John Beaumont; *http://www.beaumontinc.com/*

[8] Compiled by Jagroop Singh

Additionally, I gratefully acknowledge help from Jagroop Singh and Mukund Patankar in some of the discussion in this chapter.

5 Processing

Introduction: Injection Molding

Injection molding makes possible the mass production of intricate, high performance, tight tolerance parts with very little secondary labor and minimal waste. Continuing the theme for a holistic approach, let us proceed to the final wheel–injection molding.

As before, the design can significantly affect the processing and cost, and vice versa.

5.1 The Molding Machine

Figure 5.1 shows the schematic diagram for the basic components of an injection molding machine.

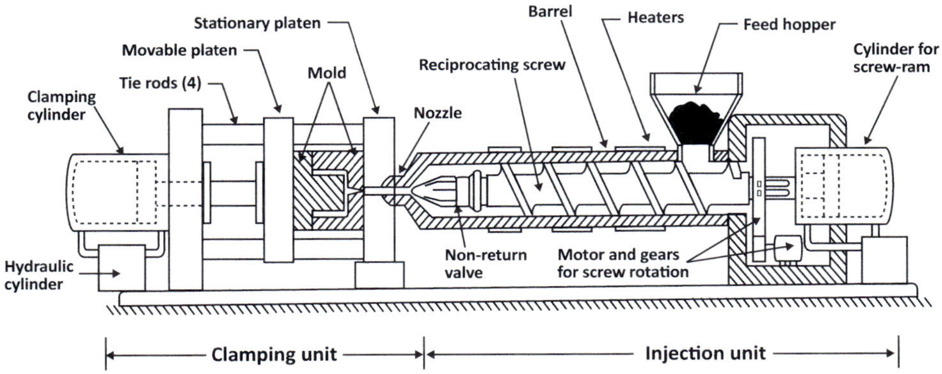

Figure 5.1 Injection molding machine schematic diagram [1]

5.1.1 Injection Molding Machine Components

The basic components are:

- The feed system, consisting of
  - The feed hopper
  - The injection barrel or cylinder, screw, and nozzle
  - The barrel and nozzle heating system
  - The screw and barrel rotating system and translation system, which can be hydraulic or electric
- The clamping system, consisting of
  - The fixed platen on the injection side
  - The moving platen on the clamp side, which can be
    - fully hydraulic
    - toggle operated
    - electric operated
    - or any combination of the above
- The power system, which can be hydraulic, electric, or hybrid
- The support structure
- Controls

There can be more than one feed system for two or more shot molding.

We will not go into greater details of the molding machine other than the feed system, as they are purely "mechanical" and do not directly affect the properties or performance of the parts, as long as they have the required power and modern controls for scientific molding, as discussed later in this chapter. Suggested sites will be provided in the reference section of the book (Chapter 9), for those with more interest.

From a purely plastics point of view, the feed system—which includes material drying, heating, plasticizing, and conveying from the hopper to the mold—is an area of special concern, as it can affect the properties and performance of the parts.

5.1.2 The Feed System

Figure 5.2 top shows the cut section through the barrel at the start of the very first cycle. This is the only time when the barrel is not completely filled with plastic. (The barrels are shown partially empty here for clarity.) The screw is at its furthest position and is made to rotate by the power system (hydraulic or electric). As it rotates, the material, in form of pellets, fills the empty spaces in the screw. The pellets also get heated or plasticized by the shearing action against each other and the steel walls. The shear heating is supplemented by heating coils around the barrel. It is important to remember that the majority of the thermal energy to plasticize the plastic comes from the shearing action. The main function of the heater band is to control the temperature and provide the supplemental heat needed for the control. A thermocouple attached to the barrel regulates the heat supplied by the bands.

Looking at Figure 5.2 middle, the material keeps moving to and collecting the front of the screw–in the process the screw moves backward. A non-return valve in the front prevents the material from flowing back into the screw. Controlled "back pressure" on the screw from moving back helps create the extra shear needed to fully plasticize the pellets and homogenize the "melt." Back pressure is an important part of the processing set up. More on the subject later.

Once full, the hydraulic or electric system pushes the screw forward. Because of the non-return valve, the screw now acts as a piston. Figure 5.2 bottom shows the screw at the end of injection.

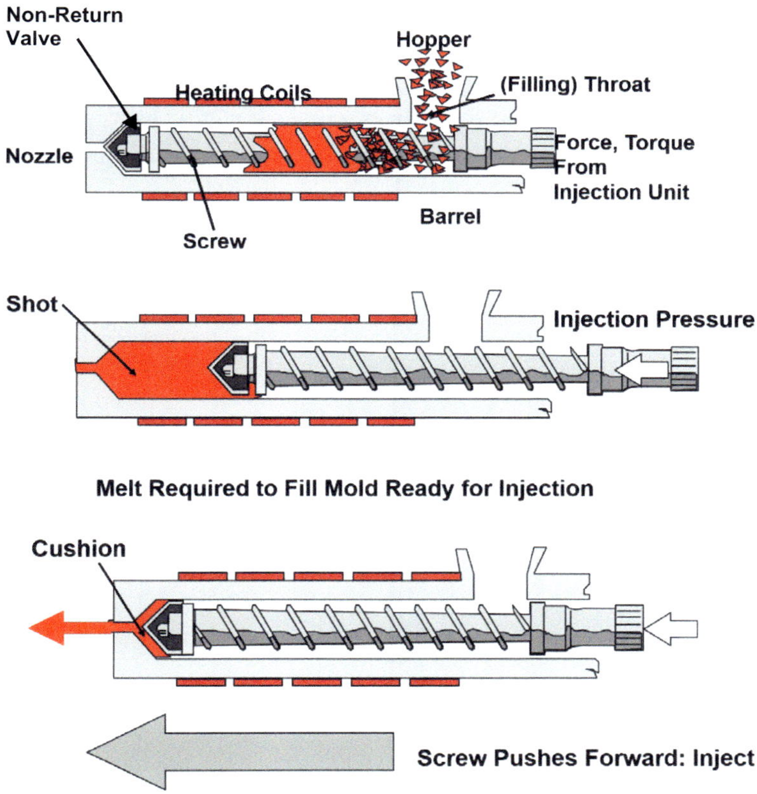

Figure 5.2 The feed unit and the injection process [2]

A small amount of material, called "cushion," is left in front of the screw. The cushion has two functions:

- Make sure the part is completely packed in every shot
- Prevent damage to the screw and cylinder by making sure that the former does not strike the latter. A leaky valve may cause the cushion to flow back into the barrel, causing shot to shot variation in the filling of the part.

The screw rotation is started for the next shot once the gate on the part is frozen and no more material can be pushed into the mold. Once the screw has moved all the way back, it is ready for the next cycle.

5.1.2.1 Screw Details

The screw consists of three zones:

1. The feed zone
2. The transition zone, and
3. The metering zone

The feed zone picks up the pellets from the throat and transfers them to the next zone. In the transition zone, the channel depth is reduced to compress and shear the plastic and to eliminate the air pockets. The final zone is the metering zone, which further shears and brings the temperature of the pellets to the processing range.

Screws have two specifications: L/D and compression ratios. The L/D ratio is the length divided by the diameter of the screw. One of the more common L/D ratios is 20. The compression ratio is the ratio of the height of the feed zone to that of the melting zone (h_f/h_m). In the case shown in Figure 5.3, the ratio is 2:1, which is a common ratio. (Ideally, every material has its own screw ratios to optimize the plasticizing. However, from a practical point of view, most machines in any shop have a general purpose ratio.)

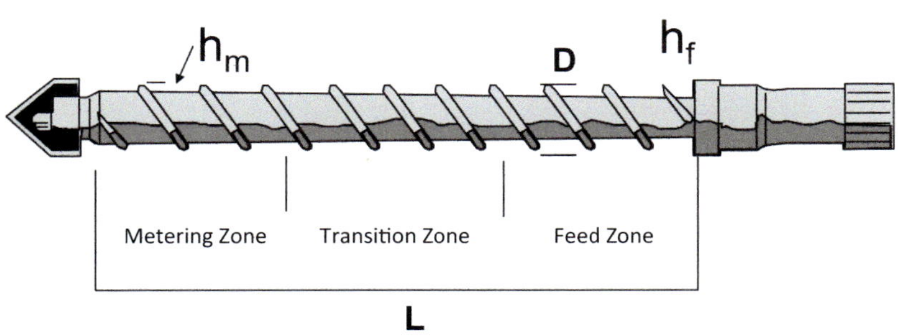

Figure 5.3 Typical cross section of a screw [2]

Figure 5.4 shows an actual screw and check ring valve. The close-up of the valve open (screw moving back to fill barrel) is in the middle and closed (screw moving forward to inject) is at the bottom.

Figure 5.4 Actual screw and check return valve

5.1.2.2 Clamp Tonnage and Capacity

Injection molding machines are specified by their clamp tonnage and capacity. The clamp tonnage is the maximum clamp pressure that the machine can exert on the mold in order to prevent its opening under the plastic pressure. It is generally specified in tons.

The capacity is the volume of the barrel in the amount of plastic it can hold, in ounces or grams. The material used to calculate the capacity is general purpose polystyrene. Thus, if the capacity of the machine is 10 ounces for polystyrene with a density of approximately 1 g/cc, the same machine will have the capacity of 10 multiplied by 1.2 for a material such as polycarbonate with a density of 1.2 g/cc, or approximately 12 ounces.

The rule of thumb for tonnage required for plastics such as polycarbonate is 4 tons per square inch or 6.45 cm^2 of projected area. The projected area is defined as the total area of the part and the runners at the parting line. It will be less for easier flowing materials and more for harder flowing materials.

Please note that this is a very general approximation. The actual tonnage also depends on the part thickness and gate placement, among other factors. Suhas Kulkarni, the author of *Robust Process Development and Scientific Molding* [6], provides an at length discussion on this subject and more information is provided in Chapter 9.

Figure 5.5 shows a cup being placed in two different ways at the parting line. The projected area will be different in both cases. In the case of the image to the right, the projected area will be the larger diameter of the cup. The diameter in this case is approximately 7.5 cm, so the projected area will be approximately 44 cm^2. The tonnage required, therefore, will be $44/2.5/2.5 \times 4$, equal to approximately 28 tons. If the cup is placed as shown in the middle image, the area is a trapezoid with sides of 7.5 cm and 5 cm and a length of 10 cm. The projected area in this case is approximately 62.5 cm^2. The tonnage required then is $62.5/2.5/2.5 \times 4 = 40$ tons.

In both instances, the area of the runner, gate, and sprue has to be added to calculate the total projected area.

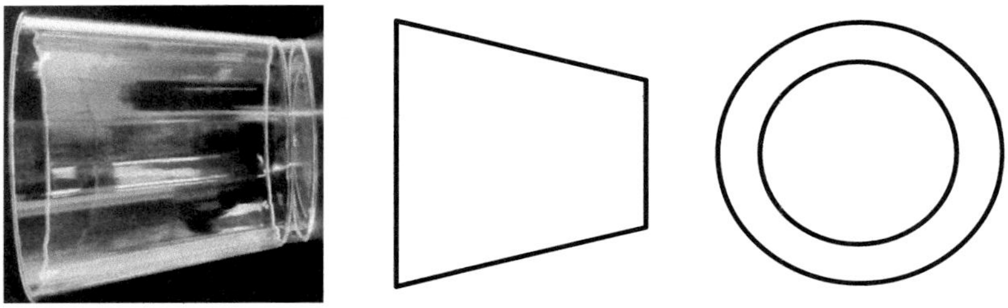

Figure 5.5 Two different ways of placing a cup at the parting line

5.1.2.3 Intensification Ratio

Another important machine specification is the intensification ratio. This is the ratio of the piston area of the hydraulic system of the machine to the area of the barrel. It is a common mistake to assume this ratio as 10, which used to be one of the common ratios in older machines.

This is a very important number and the basis for setting the hydraulic pressures on the machine. *The plastic pressure, and only the plastic pressure, is important for setting the processing conditions, and it is equal to the intensification ratio multiplied by the hydraulic pressure.*

Intensification ratio = area of the piston/area of the screw. Simplifying the constants, this can be reduced to: D^2/d^2, where D and d are the dimensions of these two areas.

Figure 5.6 shows a typical hydraulic cylinder and screw barrel arrangement. In this case, if D is 90 mm and d is 30 mm, the intensification ratio is $90^2/30^2 = 9$.

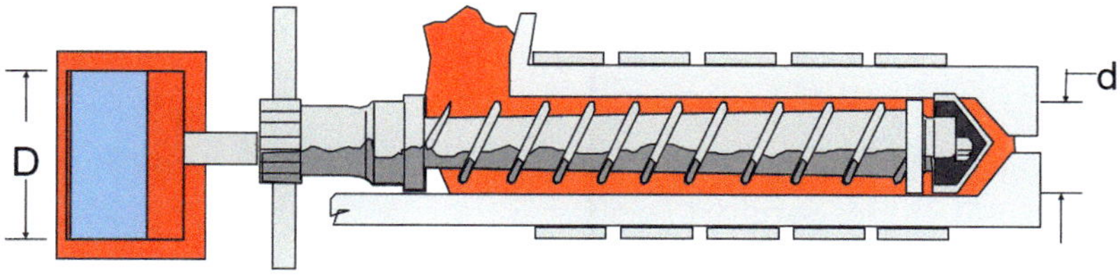

Figure 5.6 Intensification ratio [2]

5.2 The Injection Molding Cycle

Figure 5.7 illustrates a typical molding cycle, which consists of the following components:

- Mold close
- Injection
- Packing
- Cooling
- Mold open and eject

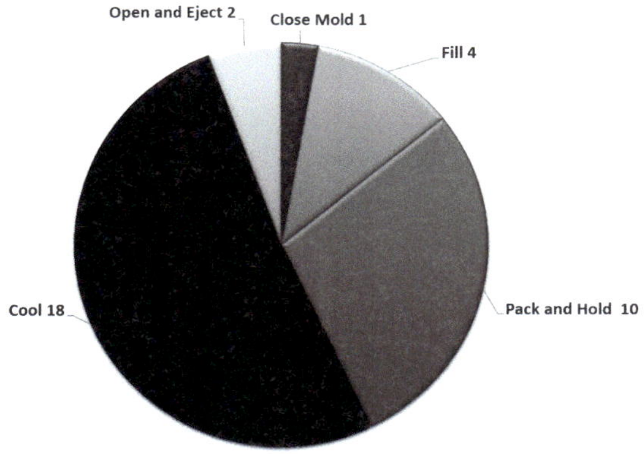

Figure 5.7 A typical injection molding cycle in seconds

Now let us look at what happens during the various stages of the molding cycle. Figure 5.8 shows a simplified view of the plastic or hydraulic pressure during the various stages. Figure 5.9 shows a representative trace of the pressure in the actual molding cycle.

5.2.1 Filling or Velocity Stage

The first stage is called the filling stage. The important factor here is the velocity of the screw moving forward. In this stage, the system provides as much pressure as needed to maintain the required velocity. Normally, the part gets about 95% or a little more filled during this stage.

5.2.2 Packing

The second stage is where a constant pressure is maintained. The pressure is somewhat less than the peak pressure in the velocity stage. The part gets fully packed in this stage. The pressure is also maintained to account for the shrinkage of the part. Once the gate "freezes," the pressure is turned off. The screw now starts its rotation to fill up for the next shot. A controlled amount of resistance to the backward motion called back pressure is maintained on the screw to help homogenize the material in the barrel and to supplement the shear required to plasticize the next batch of material.

5.2.3 Cooling and Ejection

After allowing a controlled amount of time required to cool the parts sufficiently for ejection, the moving half of the mold moves backward and the ejection system moves forward to eject the part. The removal of the part from the ejector system may be due to gravity or operator/robot-assisted. The mold is now ready to close for the next shot and the cycle repeats.

The detailed requirements on the pressure, velocity, and time will be discussed in a later section of this chapter.

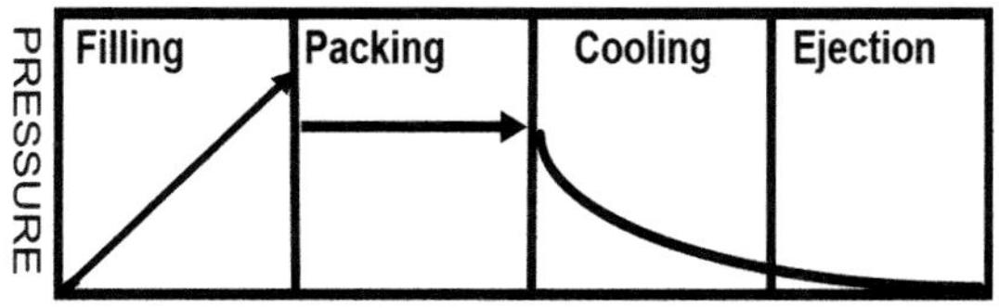

Figure 5.8 Simplified pressure history during molding cycle [2]

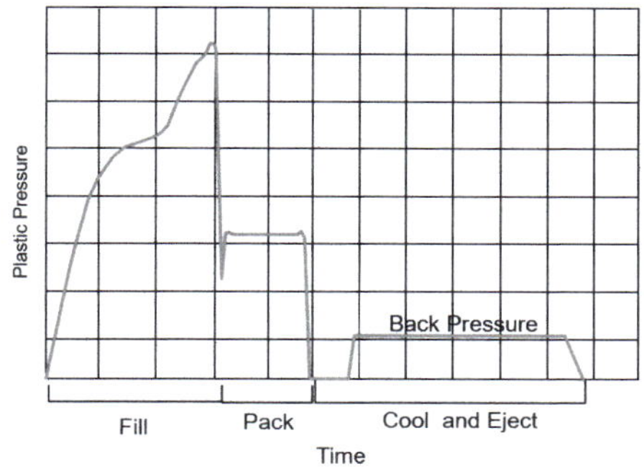

Figure 5.9 Representative plastic pressure trace during the molding cycle [2]

5.3 Process Capability

For details on process capability, please see Chapter 8.

5.4 Basic Controls in Injection Molding

It is essential that the designer be aware of the basic processing controls that need to be in place to optimize the ultimate performance of the product.

5.4.1 Drying Time, Temperature, and Drying of Air

5.4.1.1 Background

As discussed in the materials chapter (Chapter 2), the longer the AMW (average molecular weight), the higher the impact strength of a given plastic.

Before we go into the details of drying, let us look at the two types of polymers in terms of synthesis. These are condensation polymers and addition polymers.

Condensation polymers are produced by a process by which two molecules join together giving up a small molecule, often water. Common examples include polycarbonates and nylons.

Addition polymers are formed by the joining of molecules by sharing their bonds in the reaction process known as addition polymerization. Common examples include polyethylene, polypropylene, PVC, etc.

Many of these are hygroscopic and attract molecules of water in the air, which attach themselves to the polymers. This is much like common salt attracting moisture and "lumping up" in a humid environment. Going back to high school chemistry, the formula for this mixture is:

$NaCl{\cdot}nH_2O$

and it looks like Figure 5.10.

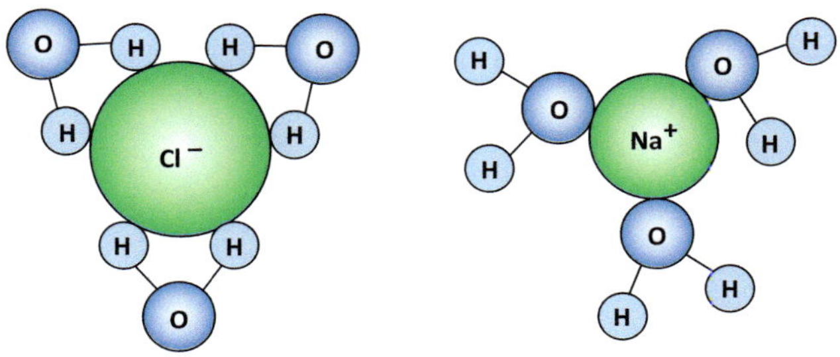

Figure 5.10 "Wet" salt molecules

Also remembering from high school chemistry, these are mixtures, not compounds, because there is no chemical reaction.

In both types of hygroscopic polymers, if the water is not evaporated away prior to molding, surface defects such as splay may occur. In the case of addition polymers, there is no significant effect on the processing or physical properties due to this small amount of moisture. In the case of condensation polymers, however, the effect on the processing and physical properties will be substantial. Before we go into the reason why, let us look at a typical condensation reaction.

Figure 5.11 shows the reaction for polycarbonate. During the reaction, bisphenol A and carbolic acid give up a hydroxyl group and a hydrogen atom as water molecules, respectively, to chemically join together. This reaction is continued until the chain grows to the required molecular weight–in this case typically about 50,000.

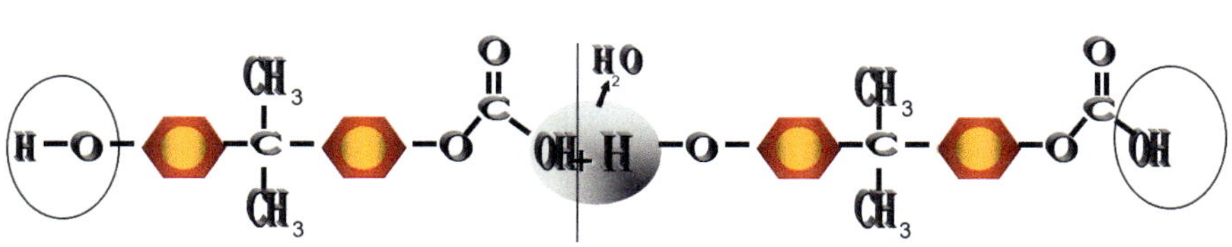

Figure 5.11 Condensation polymerization of polycarbonate (PC) via condensation of water

The reverse of this reaction is the mechanism by which PC can degrade in the presence of moisture and high heat and pressure associated with injection molding.

The typical average molecular weight of polycarbonate is 50,000. One molecule of water will split into two chains of 25,000 AMW each!

In terms of processing, this reduction in molecular chain length will dramatically vary the viscosity and, therefore, the shot to shot consistency–resulting in parts with flash to underfilled ones. In terms of physical properties, the impact strength will be significantly reduced.

5.4.1.2 The Drying Process

There are two types of drying processes: ambient air drying and desiccant drying.

Addition polymers can be dried adequately by circulating heated ambient air around the pellets at the right temperature and for the right amount of time.

In the case of condensation polymers, the ambient air is too "wet" to drive off the water molecules adequately. In this case, the air has to be dried by passing it through desiccants. In addition to temperature and time, air dryness also has to be factored in the process. The amount of humidity in the air, as we know, is indicated by its dew point (or the temperature at which moisture in the air will change from a vapor to a liquid). The higher the moisture content, the higher the dew point.

For most resins, the dew point of the drying air needs to be −20 °C to −40 °C. Table 5.1 indicates the drying temperature and times for some common resins.

Table 5.1 Drying Temperature and Time for Some Common Resins

Resin	Drying temperature	Drying time
PPO	95 °C	1-2 hours
ABS	80 °C	2-4 hours
PC	120 °C	3-4 hours
Polyetherimide	150 °C	3-4 hours
PBT	120 °C	2-4 hours

Figure 5.12 shows the arrangement for a typical desiccant drying system. Shop air is made to continuously pass through hot desiccant beds and it is heated to the required temperature. This hot, dry air enters the hopper containing the pellets at the bottom and exits at the top, carrying in the process the moisture from the pellets along with it. The air is then cooled in an aftercooler and enters the desiccant bed again. There are generally two or more desiccant chambers. Once one chamber gets saturated with moisture, a new one is rotated into its place. The saturated chamber is "recharged" by heating it to drive off the moisture, and the process continues. The temperature for recharging the desiccant bed is substantially higher than the one for heating the drying air. As one can see from the figure, they are 550 °F and 260 °F, respectively, in this case.

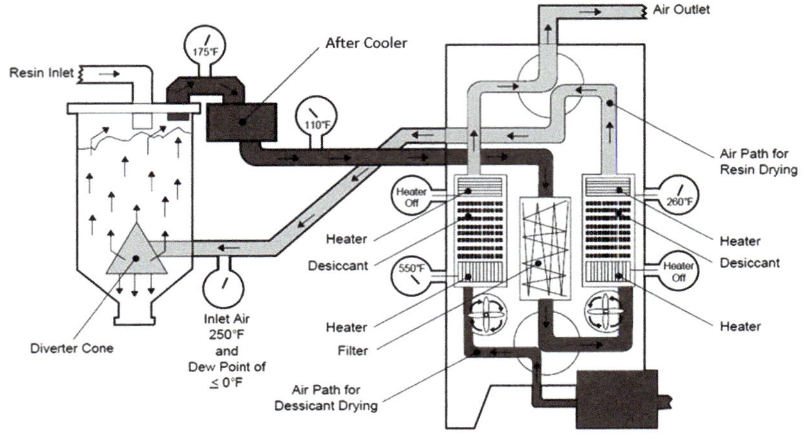

Figure 5.12 The drying system [15]

Figure 5.13 Typical floor mounted desiccant dryer [16]

The dryers can be mounted directly onto the molding machines or be placed on the floor. Figure 5.13 shows a typical floor mounted dryer in use in a molding shop. The drying air supply hose is coming out from the left side of the dryer and the return hose is going into the right. The hoses are connected to the hopper (shown to the left) at the bottom and top, respectively. On the top of the hopper there is a pump that can supply dried resin to one or more molding machines. There can also be a central system that supplies dried material or dried air to multiple machines.

5.4.1.3 Control of Drying Time

This is critical. As indicated in Table 5.1, it requires a minimum amount of drying time to get the resins to the maximum recommended moisture content for various resins, as shown in Table 5.2. The hopper volume, therefore, needs to be matched to the shot size and cycle time. As an example, for a PC part, if the cycle time is 30 seconds and the shot size is 60 grams, for a minimum drying time of 3 hours (180 minutes), the hopper has to have a capacity of 180 × 60/30 × 60 grams, equal to 21.6 kg.

There is also a maximum drying time to prevent possible degradation of the material exposed to continued high temperature. Consult with the resin supplier, for this is not readily available information, and make sure the hopper is not oversized.

Table 5.2 Recommended Maximum Moisture Contents for Common Resins

Acrylic	0.02-0.08 %
ABS	0.02-0.08 %
SAN	0.02-0.08 %
PET	0.02 %
Polycarbonate	0.01-0.02 %
Nylon 66	0.08-0.15 %

5.4.1.4 Control of Dryness

This is the area where most of the molders fail due to inadequate monitoring and inadequate equipment and/or faulty equipment, such as a drying air hose that has a leak. The most common monitoring equipment is a dew point meter, which can be self-standing or mounted onto the system for continuous monitoring. If a standalone meter is used, the dryness should be measured and recorded on a regularly scheduled interval.

In my opinion, the commonly available moisture analyzers are time-consuming and not accurate enough for the low levels of moisture required.

In any event, moisture analysis is an after the fact monitoring and a substantial amount of parts may have been produced by the time it is determined that the moisture content was too high.

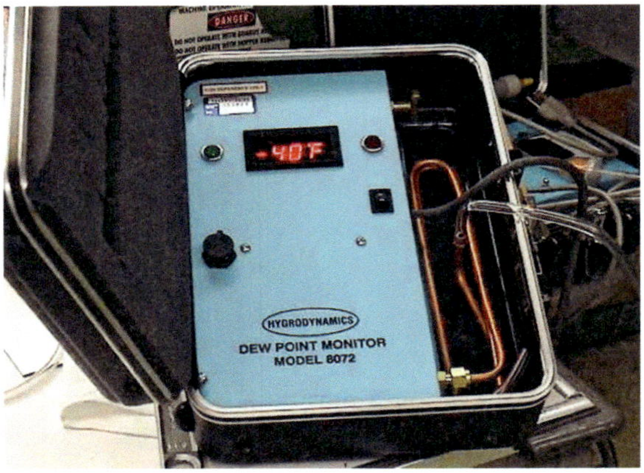

Figure 5.14 A typical dew point meter

5.4.2 Melt Temperature

As discussed earlier, most of the thermal energy needed to raise the temperature of the melt comes from the shearing action and not the heater bands around the barrel. Let us keep this in mind throughout the discussion on controlling the melt temperature.

Within a specified range, the flow length increases with the rise of the melt temperature, as shown in Figure 5.15 for PC, SABIC Lexan 101. For a wall thickness of 0.120 in (3 mm) the flow length changes from about 13 inches to about 17 inches when the temperature is increased from 550 °F to 600 °F.

Caution: The melt temperature is not a variable that should be used for controlling the flow length and, more specifically, to get the "last oomph" for difficult-to-fill parts. Ideally, the temperature should be controlled to the nominal temperature recommended by the material supplier for optimum process capability. As an example, the specified melt temperature for Lexan 101 is 290-320 °C, so the target temperature should be 305 °C.

Some of the important reasons for this are:

- Prevent impact on the process capability (Cpk)
- Prevent degradation of the polymer (see Chapter 2, Section 2.2.6.1)
- Prevent degradation of the lower molecular weight additives (pigments, flame retardants, etc.)
- Prevent increased cost due to longer cooling time needed

If there are issues with inconsistent filling of the part, basic changes may be needed on the part design or the runner system.

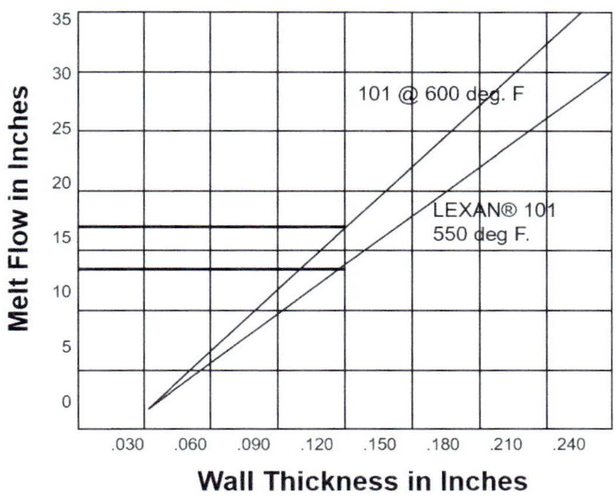

Figure 5.15 A typical flow length based on the melt temperature [2]

5.4.3 Mold Temperature

In its most basic form, the mold temperature controls the rate at which the plastic cools. A majority of it is achieved in the mold before ejection, in a relatively short time, and the remainder of the cooling is done at the ambient temperature outside the mold. As an example, the average part temperature at ejection for polycarbonate may be 130 °C. The rest of the cooling is done at room temperature and it may take many hours for the innermost layers to get to the ambient temperature.

Higher mold temperatures will result in lower molded-in stresses but will also result in a longer cooling and cycle time, and vice versa. Thus, the mold temperature is set based on the best compromise between molded-in stresses and the cost of a longer cooling time. As an example, optical quality lenses are molded in extremely hot molds with a very long cooling cycle, to minimize molded-in stresses and therefore the birefringence (see discussion in Section 2.4.2.3.3, Chapter 2). On the other hand, cold drinking cups made out of clear polystyrene are molded in very cold molds with very fast cycles for the reason of cost. The high orientation and molded-in stresses in the cup are not detrimental to its functional requirement.

In the case of semicrystalline materials, the mold temperature will also dictate the amount of crystallinity developed within the mold. If the mold temperature is too low, the crystallinity will be incomplete and continue to develop outside the mold, leading to part distortion and reduced overall strength.

For the same reason as with the melt, the mold temperature should be set at the midpoint of the mold temperature recommended by the resin supplier. Using the Lexan 101 example again, SABIC recommends 80-120 °C. Ideally, then, the mold should be run at 100 °C.

Caution: The mold temperature settings should not be used to reduce surface sink marks, warpage, reduce cycle time, help with ejection, change the dimension of the part to match specifications, etc. Mold temperatures at the extreme of the range or beyond will reduce the performance of the part and overall process capability. The molding issues should be fixed by fixing the part or tool design.

5.4.4 Fill Time and Shear Rate

Fill time is the time required to fill the cavity approximately 95 % full. Before we go into the fill time discussion, it is important to review the fundamentals of viscosity. Going back to the basics, viscosity is the quantity that describes the resistance of the fluid to flow. Fluids resist the motion of layers with different velocity contained within them.

There are two types of fluids: Newtonian and non-Newtonian. Newtonian fluids are the ones whose viscosity is independent of the shear rate or the rate at which they are flowing through a certain opening. Water is the most common Newtonian fluid. If one is pumping water through a hose, the viscosity does not change whether the water is just dripping or being forced to flow at a high rate when the faucet is opened all the way up. Non-Newtonian fluids, on the other hand, are the ones whose viscosity changes based on the shear rate or the rate at which they are being made to flow through an opening.

Polymers are in the non-Newtonian family. Therefore, the fill time will decide how viscous the polymer is in the process of filling the cavity. A short fill time will increase the shear rate and the viscosity will drop. A longer fill time means that the shear rate is low and therefore the plastic will be more viscous.

Figure 5.16 shows the typical curve for the fill time against the relative viscosity. The term "relative viscosity" here is used just to show the relative ease of flow of the plastic in the given situation. A higher number means the plastic is more difficult to flow and vice versa.

Looking at the curve, the relative viscosity is extremely high at a fill time of 10.69 seconds. The viscosity drops down to a very low value when the injection time is reduced to 0.1 seconds. The shear rate is the reciprocal of the injection time (as an example, at a 0.1 seconds injection time, the shear rate is $1/0.1 = 10.00\ s^{-1}$). Therefore, the shear rate is higher for a lower injection time and vice versa.

The section on scientific molding in this chapter will discuss the effect of changing the fill time in much greater detail. For now, suffice it to say that even small changes in the fill time will change the viscosity of the plastic leading into parts that range from being overpacked to short shots.

In addition to the variation in the filling of the parts, the variation in the shear rate will cause different behaviors in the polymer. Low shear rates will not completely homogenize the plastic and therefore cause inconsistencies in the properties of the part. Very high shear rates may degrade the additives in the plastic, in addition to the polymer itself. Most suppliers provide a range of useful shear rates for their resins.

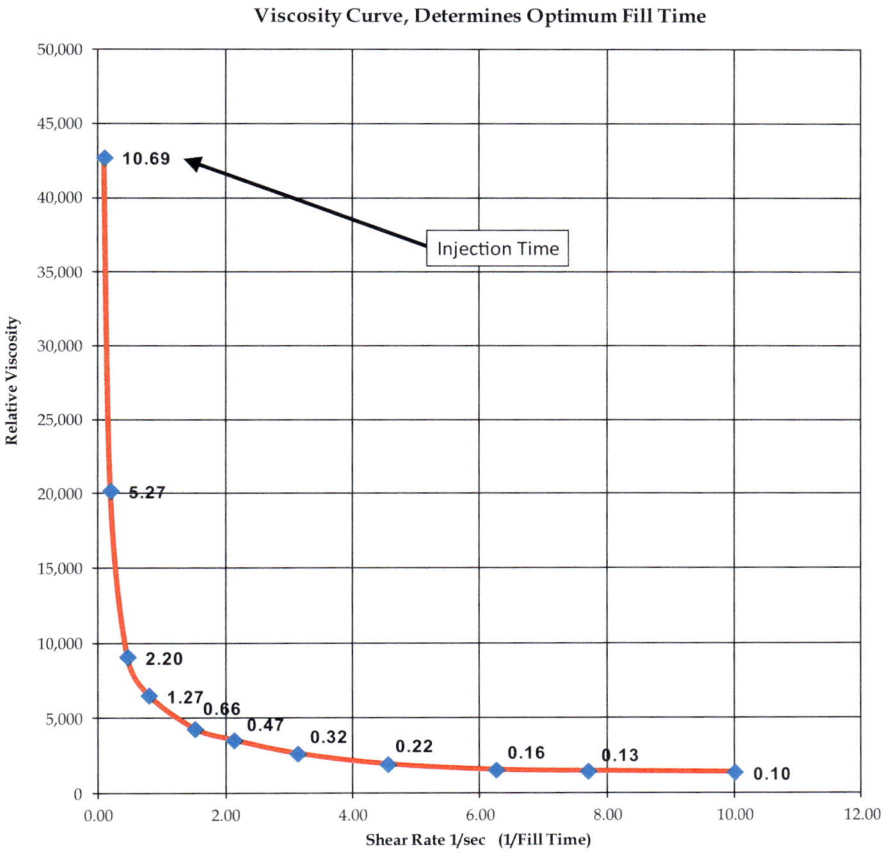

Figure 5.16 Typical relative viscosity versus shear rate (1/fill time)

5.4.5 Packing Pressure and Time

The next stage in the molding cycle is the packing phase. The injection phase fills the part approximately 95% full. The packing phase fills the part completely and compensates for the shrinkage of the plastic during the injection and packing phases. In general, the packing pressure may be about 80% of the injection pressure. The packing time depends on the time it takes to freeze the gate. If the packing pressure is withdrawn before the gate is frozen, the material may flow back into the barrel and cause the part to be underpacked. Continuing the packing after the gate is frozen is a waste of energy, since no material can enter the cavity because of the frozen state of the gate.

5.4.6 Gate Freeze Time

The gate freeze time is the time it takes to get the temperature of the plastic at the gate to a point where it cannot flow anymore. The section on scientific molding will describe how it is measured. It is important to determine the gate freeze time in order to make sure the packing time is not less than this, as described in the previous discussion.

5.4.7 Cooling Time

The cooling time starts at injection and continues after the packing pressure has been taken off. This is the time required for the parts to cool down to the point where they can be ejected without distortion. In the case of semicrystalline plastics, this time is also important to allow for adequate crystallinity to develop.

For a given set of mold and melt temperatures and gate configuration, the cooling time remains constant from one set up to the next.

5.4.8 Residence Time

The residence time is the time that the plastic remains in the barrel from its entry into the throat to its discharge from the nozzle. Resins will start to degrade if kept close to the processing temperature for more than a certain time. Here is how to calculate the residence time:

- Barrel capacity: b oz or approximately $(30 \times b)$ g
- Part weight: p oz or approximately $(p \times 30)$ g
- Runner and sprue weight: r oz or approximately $(r \times 30)$ g
- Density of material: d
- Cycle time: c seconds

Residence time in minutes = $d \times b/(p + r) \times c/60$

As an example:

- Barrel capacity: 10 oz or approximately 300 g
- Part weight: 1 oz or approximately 30 g
- Runner and sprue weight: 2 g
- Cycle time: 30 seconds
- Density: 1.2 g/cm^3 (polycarbonate)

Residence time in minutes = $d \times b/(p + r) \times c/60$ or $1.2 \times 300/32 \times 30/60$ = 5.62 minutes

According to John Bozzelli, known expert in scientific molding, the process should use somewhere between 25% and 75% of the barrel or shot capacity, for the following reasons [3]:

If the shot weight is less than 25%:

- There may be color and material degradation
- Transparent materials may acquire a darker shade
- The velocity or fill time consistency may get impacted due to the relatively short stroke and the viscosity reduction due to degradation

If the shot weight is over 75%:

- There may be problems with melt uniformity. With large shots, a large portion of the feed section will wind up behind the feed throat as the pellets fill the flights, leaving only a few flights to prepare the resin for the transition or compression zone.

Effect of Residence Time and Barrel Temperature on the Properties

In addition to the obvious discoloration and "browning" of clear plastics, the combined effect of longer residence time may cause serious issues with the rise in MFI (Melt Flow Index) and the loss of impact strength of materials. Figure 5.17 shows the effect of residence time and barrel temperature on Covestro PC/ABS Bayblend. The study was conducted at the recommended pellet moisture content: wetter, and the results would have looked much worse; drier, and they would be better. At 20 minutes of residence time and the barrel temperature at 560 °F, there is a dramatic drop in impact strength.

Figure 5.18 shows the effect of residence time and barrel temperature on the same material on the melt flow rate (MFR). At only 1.7 minutes of residence time and the barrel temperature at 560 °F, there is an almost 25% increase in the MFR.

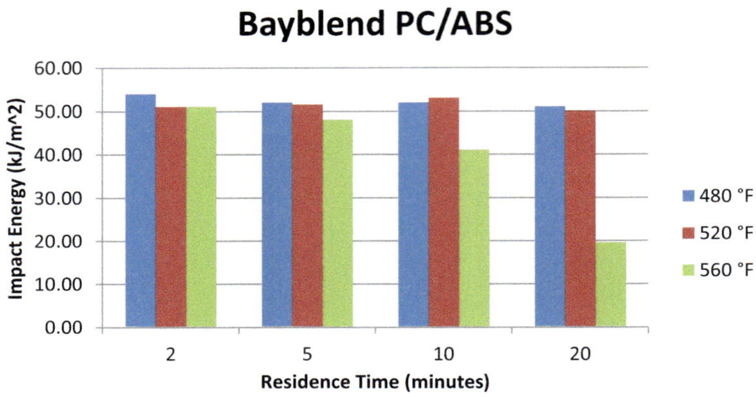

Figure 5.17 Effect of residence time and barrel temperature on impact strength [15]

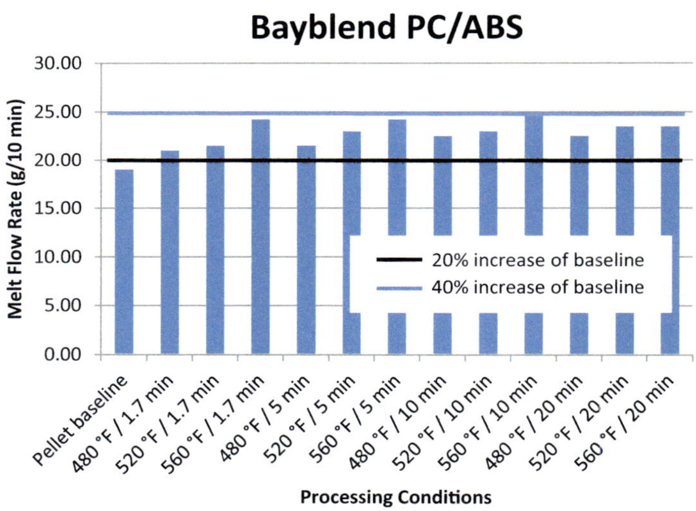

Figure 5.18 Effect of residence time and barrel temperature on melt flow rate [15]

5.5 Common Processing Issues

Processing issues may indeed be due to poor processing, or may appear to be so because of design, tooling, or material issues. They may also be a combination of all of the above. The next section discusses some of the more common processing issues.

5.5.1 Poor Drying

Every material has a specified maximum allowable moisture content based on a balance between processing complexity and performance. As an example, the maximum permissible moisture content for Bayer Makrolon PC resins is 0.01 %. As discussed in the materials chapter (Chapter 2), a higher than allowable moisture content can cause the molecular chains to break and dramatically reduce the impact strength of the molded part.

In terms of processing issues, a very high moisture content can result in visible splay. See Figure 5.19.

Figure 5.19 Visible splay in plastic part

The breaking of the chains can also reduce the viscosity in an uncontrolled way, resulting in shot to shot variation from underfilled to parts that flash.

5.5.2 Hesitation

If the plastic filling a cavity has the option of filling either a thin section or a thick section, it will tend to fill the path of least resistance to flow. This can result in plastic in the thin section stopping or slowing down significantly. By the time the thicker section is filled, the plastic in the thin section may have cooled enough to result in a no flow condition or require more pressure to fill than is available.

Hesitation can be caused by less than optimum placement of the gate, thus making a certain section of the part easier to fill than the other, or by a design issue where the thick section is surrounding a thin section.

Figure 5.20 shows hesitation due to the placement of the gate. In the left image, the plastic flows in a parallel direction to the runner. When the plastic reaches the vertical tab hanging out from the bottom, it needs to change direction to be able to fill it. It therefore fills the rest of the part and finally tries to fill the tab. By this time, the material has cooled down enough that it has difficulty filling the tab, ending in a short shot. If the gate is moved to the top and the flow into the tab is in the direction of the flow, it has no issue continuing to the end of the tab. There is no hesitation in this case.

The illustrative animation can be seen at: *https://knowledge.autodesk.com/support/moldflow-flex/learn-explore/caas/CloudHelp/cloudhelp/2016/ENU/MoldflowInsight360/files/GUID-167DA746-9E44-4F4D-96C6-0D0998C3EF97-htm.html*

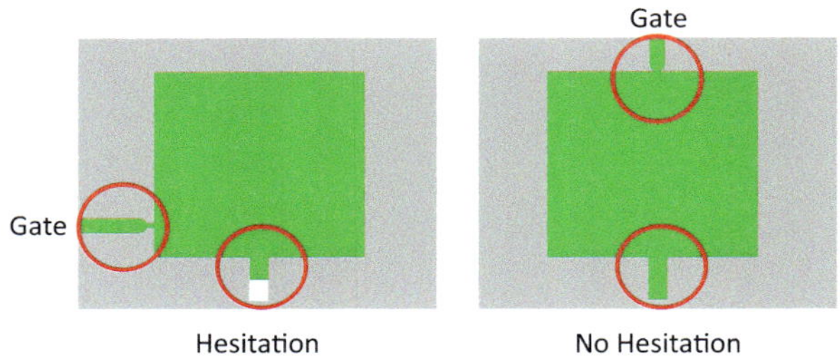

Figure 5.20 Example of hesitation due to placement of gate [4]

Figure 5.21 shows hesitation when a thin section is surrounded by a thick section. Looking at the dotted line in the top image, there is a thicker section on the outside and a thinner one on the inside. The thick section is 2.5 millimeters and the thin section is 1.5 millimeters thick. The flow goes around the thin section and fills the part to the end, leaving a void within the thin section. Once the thicker section is full, the plastic starts to fill the thin section. Due to the loss of temperature and pressure, it cannot do so fully. Even if the part gets visibly full, this section is going to be weaker than the surrounding section because it is less densely packed. It is also more susceptible to chemical attack, in addition to failing at lower stresses than the surrounding thick area. Additionally, in order to fill this thin section, higher and longer than needed pack pressures will be used. This uneven packing of the part may also cause warpage and an unnecessary energy requirement to provide the high pressure and its longer duration.

The middle image in Figure 5.21 shows the clear formation of a knit line due to hesitation. The hesitation was caused by the D shaped hole in the part. As shown in the illustration (bottom), this kind of hole resulted in a thick section adjacent to a thin section. The thick sections filled faster than the thin sections, resulting in hesitation and creating four avoidable knit lines. The part shown failed at the knit lines in use.

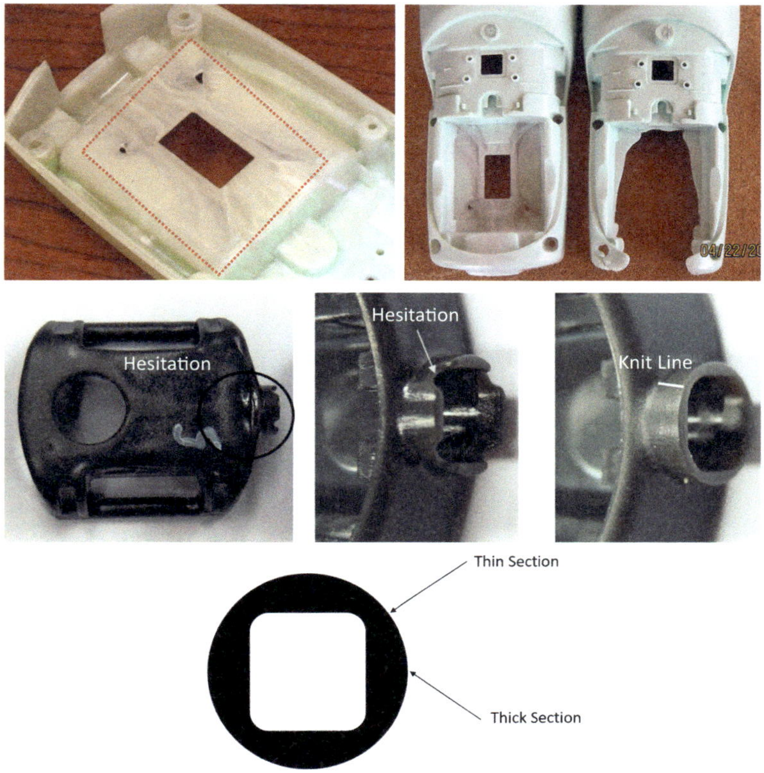

Figure 5.21 Top: example of hesitation due to thick section surrounding thin section; the short shot shows the hesitation. Middle: the rightmost image shows unnecessary knit lines being created due to hesitation in a D-shaped hole (bottom)

Figure 5.22 shows the simulation of a part that has thick and thin sections adjacent to each other. This is a plot of the flow front progression in terms of time. Red indicates 0 to 0.01 seconds and deep blue indicates about 1 second.

The gate is shown by the cone at about the 6 o'clock position. Wider bands indicate faster flow, as a longer distance has been traveled in the same amount of time, compared to a similar colored band in the area where the contours are more narrowly spaced. The area represented by the dotted lines has a thickness of 1.5 mm. The rest of the part has a thickness of 2.5 mm. The plastic flows a much longer distance in the thick area as compared to the thin area in the same one second. Therefore, material in the thin section is hesitating. In fact, the hesitation is so bad that there is an area of no fill at the top of the thin portion. Again, if this part were to be filled using a higher pressure and a longer packing time, there would almost certainly be flow marks and molded-in stresses in the thin area. The overall quality of the part would also suffer because of the very high packing in the thick areas compared to the thin areas, resulting in uneven shrinkage and possible warpage.

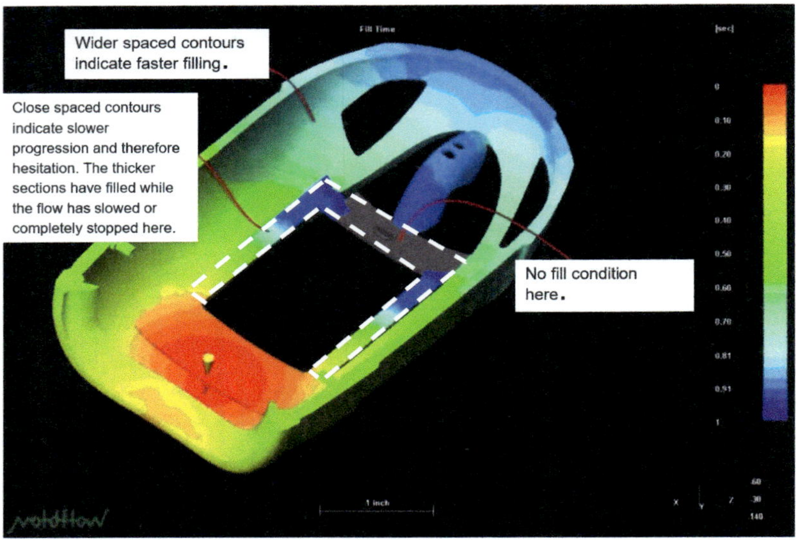

Figure 5.22 Flow simulation showing hesitation

5.5.3 Knit Line Strength and Appearance

As discussed before, knit lines will occur whenever two flow fronts meet at the end of the flow, such as in the creation of a hole in a part. Also, as discussed before, the location of the knit line can be moved either by changing the gate location or by creating flow leaders to make the flow faster in one of the directions. See Chapter 3, Section 3.2.6.2. Therefore, the creation and location of the knit line is controlled by the part and tool design.

Additionally, the material itself can affect the appearance by releasing low molecular weight volatiles from additives such as flame retardants that get trapped between the fronts and/or clog the vents around the knit lines. A term used for this is "juicing" or "plate out." See Figure 5.23. The ejector holes were completely covered with deposits from the volatiles contained in the flame retardant package in one major PC/ABS brand. The entire video can be seen at *http://www.vikpedia.org*.

Figure 5.23 Clogged ejector holes due to juicing

The appearance and strength of the knit line depend on the following:

5.5.3.1 Venting

When flow fronts come together, they generally trap air and gas between them. This trapped air and gas can affect the appearance in the following ways (see Figure 5.24):

- Not allowing the fronts to meld, thus leaving a microscopic void
- Burning the flow front due to dieseling (see Chapter 3, Figure 3.33)

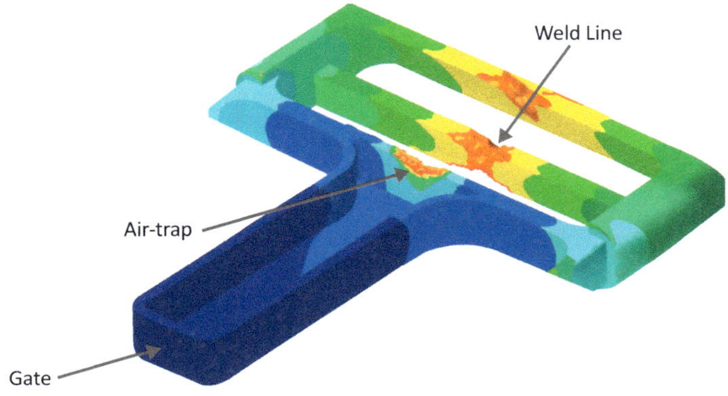

Figure 5.24 Knit line appearance due to trapped gases [5]

5.5.3.2 Melt Temperature

Higher melt temperatures intuitively should make the knit lines stronger, since the flow fronts meet at a higher temperature and therefore cause better fusion between them. However, they can also result in a higher amount of volatiles, due to the degradation of the low molecular additives and the polymer itself. These volatiles and degradation products travel faster than the polymer and therefore come in the way of an optimum fusion of the fronts. Therefore, the effect of a higher melt temperature on the weld quality depends on the specific combination of material, part, and tool design.

5.5.3.3 Injection Speed

Higher injection speeds go hand in hand with a higher melt temperature because of the increased shear. Therefore, the previous discussion also applies to injection speeds.

Additionally, on the negative side, the volatiles do not have the time to be completely vented and may get trapped between the fronts from the opposite sides, causing the weld to be weaker.

5.5.3.4 Packing Pressure

Higher packing pressures generally result in better knit lines, due to the better fusion of the flow fronts and the expelling and/or compression of the volatiles. They also help the melt flow into the nooks and crannies of the mold surface instead of just skimming it. See Figure 5.25.

5.5.3.5 Mold Temperature

The mold temperature has the most effect on the quality of the appearance and, consequentially, the strength of the knit lines. This is primarily due to two reasons:

1. The melt is kept hotter, without the degradation associated with the melt temperature and injection speed
2. The melt reproduces the surface of the mold instead of skimming along it; Figure 5.25 shows the difference in the surface with low and high pressures and melt temperatures

Figure 5.25 The left view shows a part with lower mold temperature and pressure; the right view shows the same part with higher mold temperature and pressure

5.5.3.6 Mold Surface

Textured surfaces generally hide the knit line and shiny surfaces do the opposite. Therefore, where possible, the inclusion of even a light texture improves the appearance.

5.5.3.7 Material

Finally, the material itself can add to or take away from the strength and appearance of the lines. As an example, flame retardant materials may have more volatiles that prevent fusion of the flow fronts than their non-flame retardant counterparts.

Within the same family, glossier plastics may show the knit line more than their less glossy counterparts. As an example, a straight PC may show the knit line more than a PC/ABS, which is somewhat less glossy.

5.5.3.8 Putting It All Together through a DOE

Below is a case study to validate some of the preceding discussion on the appearance of a knit line.

Figure 5.26 shows the knit lines in a PC/ABS part before and after adjusting the molding parameters. The part in question had visually unacceptable knit lines, as can be seen from the photographs.

This study is also an example on how a simple DOE (Design of Experiments–see details of the methodology in Chapter 8) can quickly and efficiently get to the best solution. Without a DOE, it might have taken days or even weeks of changing parameters one at time to come up with a solution. Even then, the solution might have not been the optimum one.

The factors considered were: mold temperature, melt temperature, and injection speed, per the preceding discussion. The mold did have adequate venting through vented ejector pins in the back.

One of the requirements of a DOE is that the effect of the study be a numerical value and not an attribute such as "good" and "bad." The task, therefore, was to convert a visual defect or lack thereof to a numerical value. In order to do this, a team of engineers and QC personnel was asked to rate the appearance of each study from 0 to 5, zero being totally unacceptable and five being totally acceptable.

The recommended melt temperature for the material was 520-540 °F, so these values were chosen as the low and high, respectively. Similarly, the mold temperatures were 175 °F and 200 °F, respectively. The injection speed values were determined from a viscosity study that will be described in the following pages, in the scientific molding section.

Looking at the DOE analysis graphs in Table 5.3, it clearly shows that the mold temperature had the greatest effect on the appearance. Notice also that the increase in injection time improved the appearance! This is contrary to intuition. Ask any molder and he/she will say that increasing the injection time will degrade the quality of the knit line, as the material fronts will be cooler when they meet. This shows the value of letting the data drive our decision against the opinions of so-called experts.

One final point. Looking at Table 5.3, the best knit line would have been at the highest mold, melt, and injection time. However, keeping these extreme settings would have reduced the "processing window," resulting in a relatively unstable process and reducing the process capability or Cpk. There will be more discussion on the processing window later in this chapter. The charts in Figure 5.27 show the above graphically.

The final settings were 530 °F, 190 °F, and 3.5 seconds for melt temperature, mold temperature, and injection time, respectively, for an appearance index of >3, which was visually acceptable as the best compromise.

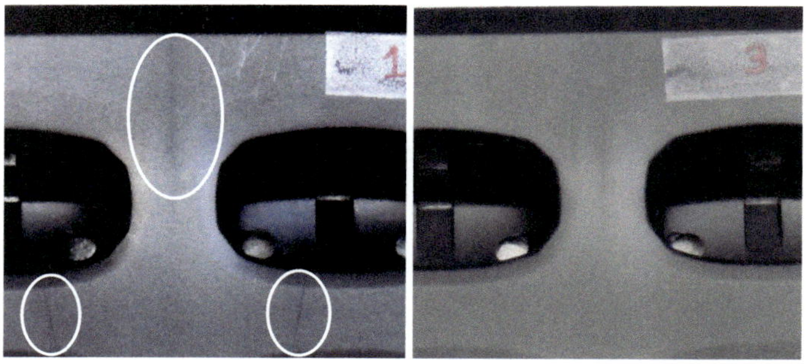

Figure 5.26 Before and after appearance of knit lines

Table 5.3 DOE Results

Experiment. No	Melt Temp F	Mold Temp. F	Injection Time Seconds	Weld Line Appearance Index
1	520	175	2	1.3
2	520	175	5	1.68
3	520	200	5	3.66
4	520	200	2	3.1
5	540	175	2	1.9
6	540	175	5	1.94
7	540	200	5	4.06
8	540	200	2	3.26

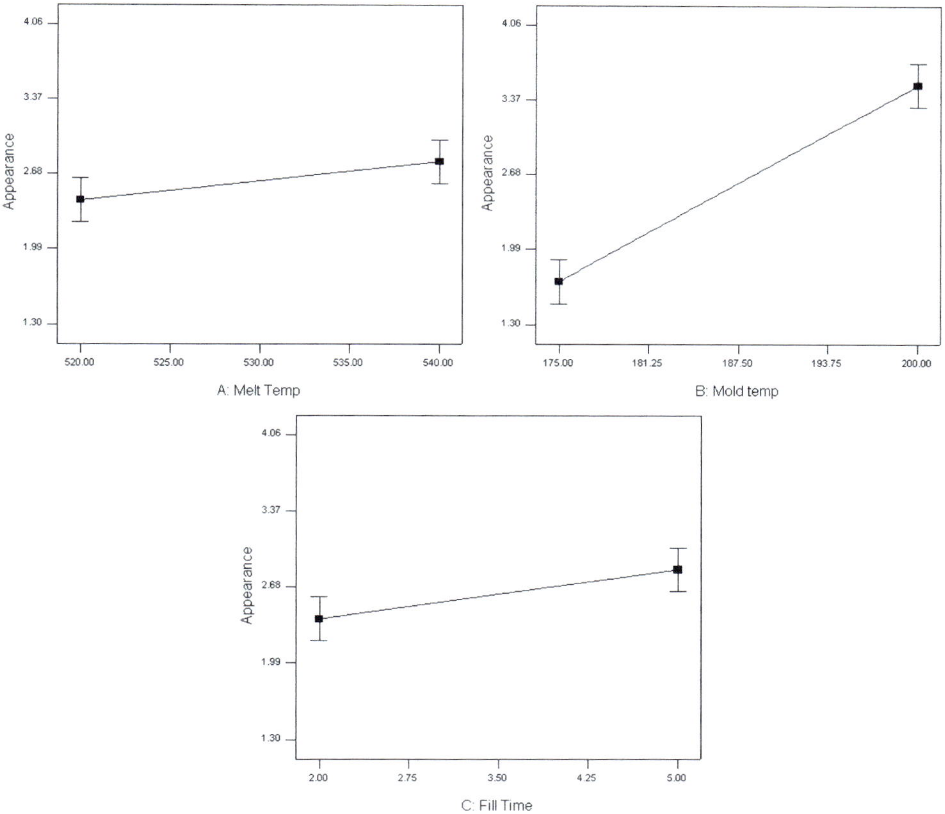

Figure 5.27 Effect of changing processing conditions on the appearance of the knit line

5.5.4 Jetting

When the material enters the mold through a properly designed and located gate, it starts a laminar flow by immediately expanding and filling up the walls above and below the gate. See fountain flow discussion in Chapter 3, Section 3.2.6.1.2.1. The outside walls freeze against the walls and the material continues to flow between the frozen layers to the end of the part.

If the gate is too small (see discussion in Chapter 4) and/or the material flows directly from the main runner to the part, the incoming material shoots past the walls adjacent to the gate, forming a snake-like structure that is apparent even after the part has filled completely. This is not just a cosmetic defect, as it causes discontinuities in the walls, resulting in lower mechanical properties. Figure 5.28 shows a properly formed laminar flow on the bottom and jetting on the top. Figure 5.29 top shows the initial creation of the jetting.

Slowing down the injection can sometimes eliminate the jetting. See Figure 5.29 middle. The left view shows the jetting. The right view shows the jetting eliminated by slowing down the injection speed. However, the processing robustness is reduced, as discussed in the scientific molding section.

The bottom view of Figure 5.29 shows a part with unavoidable jetting. The best solution to this issue is to either relocate or increase the size of the gate to the proper dimensions. The flow can also be slowed down by changing the direction of the flow as shown in Figure 5.30.

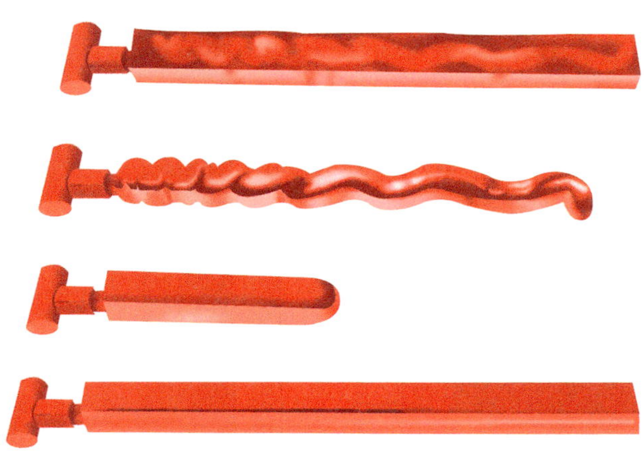

Figure 5.28 Proper flow and jetting

Figure 5.29 Jetting examples

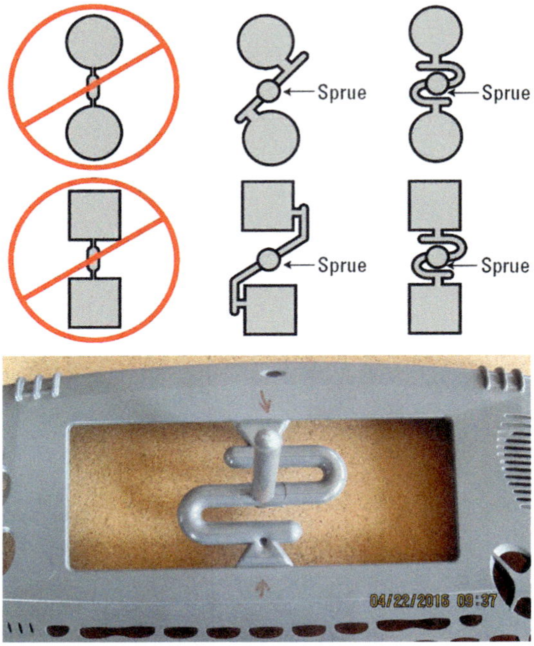

Figure 5.30 Slowing down the flow into the cavity by changing the direction of the flow (top view [15])

5.5.5 Poor Venting

As described in Chapters 2 and 4, proper venting is an important requirement both in tooling and molding. Even if the tool is properly vented to start with, proper and continuous cleaning of the mold has to be ensured. As described in the previous chapters, the lack of venting can cause short shots and burn marks due to dieseling. More serious issues such as cracking of the features at the end of flow like bosses can also occur due to poor venting. Figure 5.31 shows a lateral air trap and Figure 5.32 a perpendicular air trap, both resulting in fractures.

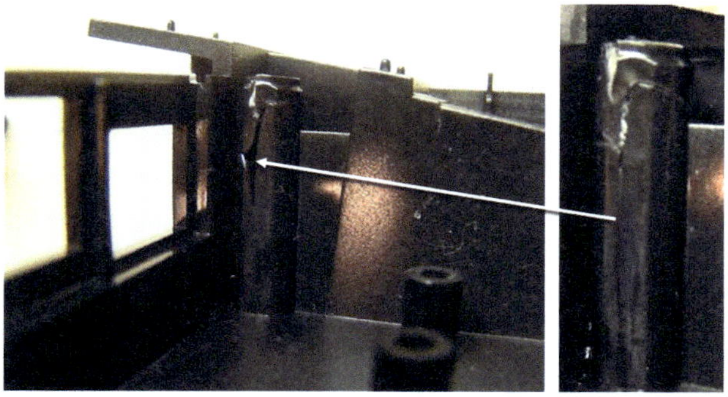

Figure 5.31 Voids in boss due to trapped gases resulting in crack in the boss

Figure 5.32 Cracked boss due to trapped gases

5.5.6 Poorly Designed Gates, Runners, and Sprue

Gates and runners deliver the material from the main sprue to the extremities of the parts with controlled loss in pressure and material, and should easily be ejected when the mold opens. Although proper runner and gate design is the responsibility of the tool design engineer, more often than not this is not the case, and the processor has to make adjustments resulting in poor part quality, lower yield, lower process capability, and higher costs.

To start with, too small a sprue diameter will cause an unnecessarily high drop in the plastic pressure, resulting in underpacked or underfilled parts in spite of well-designed runners or gates.

Some of the common issues with the gate design are:

- Too small or gated directly into the part, causing
  - Jetting as described above
  - Inadequate filling requiring excessive pressures
  - Degradation of the material due to excessive shear

- Too large, causing
  - A very long freeze time and therefore a longer cycle time
  - Distortion of the part on ejection
  - An unacceptable gate vestige
- Unbalanced gates in a part with multiple gates, causing
  - Overpacking in some sections and underpacking in others, leading to warpage, visual defects, and loss in mechanical properties
- Wrong location
  - Gated in a thin section, requiring excessive pressures and causing unbalanced filling and warpage
  - Causing a blemish on a visible surface

Common issues with the runners are:

- Too thick, causing
  - Material wastage
  - Longer cycle time–the mold has to remain open to make sure the runners have cooled down to ejection temperature instead of the part
- Too thin, causing an unnecessary pressure drop leading to poor filling of the parts
- Restrictions on the runner, causing huge pressure drops. See Figure 5.33. The "sucker pin" used to hold the runner on the cavity side of the mold is too deep. (See Chapter 4 for a detailed discussion.) It reduces the cross section of the runner and causes the runner itself to flash while the part starves. In the case of this part, this resulted in the poor packing of a knit line, causing it to crack from the sweat on the operator's hand.

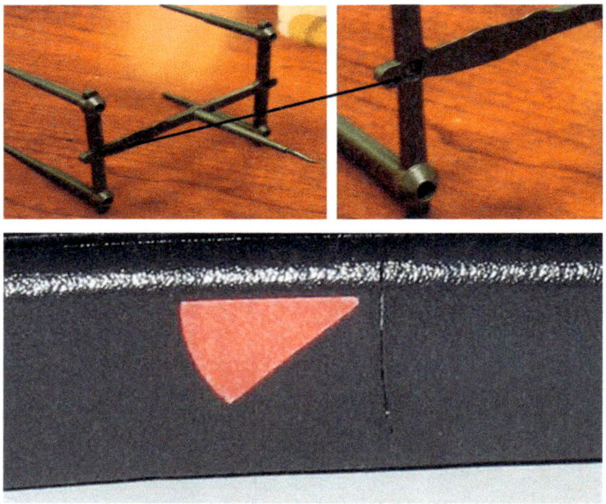

Figure 5.33 Runner showing restriction and the poor knit line condition it caused

- Unbalanced runners will cause one area of the part to fill faster than the other, causing flash on one side and underfill on the other. This may also cause the part to warp. The top pictures in Figure 5.34 show a poorly designed runner system causing the top to flash while the bottom remained unfilled. The Moldflow analysis shows the unbalance. The top is filled with a pressure of over 11,000 psi while the bottom gets only about 1,200 psi. The bottom pictures show the same with a runner system showing balanced filling in the part as well as the simulation.

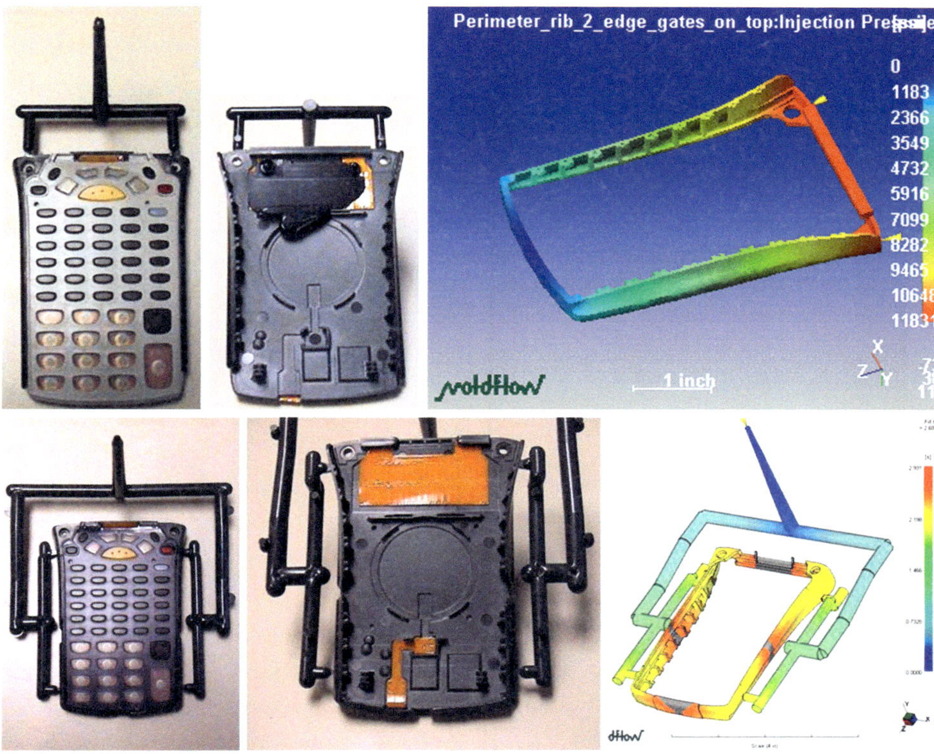

Figure 5.34 Top: unbalanced runner; bottom: after balancing

- Unbalanced filling in family molds. As discussed in the tooling chapter (Chapter 4), having a family mold that has cavities with very different volumes and/or filling pressures will continue to cause filling issues. Figure 5.35 shows a family mold with parts with different thicknesses and volumes being filled by a common runner. The top two parts have the same thickness but different lengths. The bottom part has the same length as the top one but half the thickness. The process seems to be optimized for the top part, which is filling properly. The second part is flashing because it requires less volume and the bottom part is clearly starving because it requires a much greater pressure to fill. The net result: the customer will have to accept substandard parts or the molder will have to run one part at a time, blocking the other cavities. Worse still, the molder will have to throw away three out of four cavities if he/she does not choose to block the cavities. As stated in the tooling chapter, changing the gate size on the part is not an option for two reasons:

1. There is only one gate size that will provide the optimum shear for the part
2. The slightest change in gate thickness may cause a major change in the filling pressure, thus leading to a very unstable process

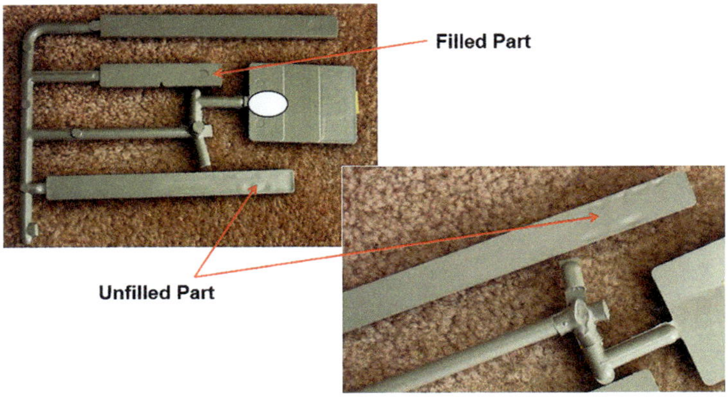

Figure 5.35 Poor choice for a family mold

5.5.7 Mold Heating

Chapter 4 has a detailed discussion on the thermal management of the mold. In a lot of cases, molders do not follow the requirements out of ignorance, poor monitoring, or intentional adjustment of the temperatures to reduce the cooling cycle or to make up for some other part deficiencies. Let us discuss these in detail:

5.5.7.1 Ignorance

In all my years of working with molders, I have found a lot of ignorance in this area. In a lot of instances, the mold cooling is based on the "seat of the pants" method. There is no monitoring of the quantity of flow to calculate the Reynolds number, the temperature difference between the incoming and outgoing water temperatures, or the relationship between the water and mold surface temperatures. Worse still, the set up person is allowed to change the water temperature at will.

5.5.7.2 Poor Controls and Monitoring

There are many ways in which the control and monitoring of the mold temperatures may be less than adequate. Some of these are:

- Faulty temperature gauges
- Clogged water lines
- Scale build up in the cooling lines
- Misconnected water hoses
- Leaking water hoses
- The actual mold surface temperature may be very different from the water temperature due to the location of the cooling lines or poor conductivity of the mold material
- Lack of formal set up documents and procedures

5.5.7.3 Intentional Overriding of Established Process Parameters

There are many reasons for this. Some of these are:

- Reducing the cycle time to improve profit by running the mold too cold (as discussed before, the cycle time is one of the most critical parameters to make or lose money on a part). Not an ethical business practice!
- Improving ejection (a cooler mold will allow easier ejection of a difficult to eject part)
- Surface quality (a cooler mold will provide a more matt surface)
- Warpage (a cooler mold may reduce the warpage–temporarily!)
- Running one half of the mold hotter than the other to compensate for warpage. Not a good idea, since different shrinkage on the opposing surfaces may cause added stresses.
- Dimensional adjustment (a part coming out of a cooler mold will shrink slightly less and may be slightly longer than one coming out of a hotter mold)
- Other make believe reasons

5.5.7.4 Effects of Running the Mold Too Hot

In terms of the physical properties, running the mold too hot (within reason) will have little effect on the properties of the part and may even improve them. The other effects may be that the part may:

- Warp more
- Be slightly shorter due to greater shrinkage
- Be a little glossier
- Hide the glass fibers in a glass-filled part leading to a resin rich surface
- Be more difficult to eject
- **Cause the cycle to be longer (a profit killer)**

5.5.7.5 Effects of Running the Mold Too Cold

This can cause many short and long term losses in properties. Some of the main ones are:

- Higher molded-in stresses due to the inability of the chains to order themselves in a random fashion
- High orientation–in a rapidly cooled part, the chains get frozen in the direction of flow
  - Refer to Figure 5.36. This is a common cold drink cup generally made from clear polystyrene. The parts are mostly molded in very cold multicavity molds to minimize the cycle time and reduce the overall cost. If one squeezes the cup in the length direction, one can see parallel cracks running down the length of the part, indicating failures between the chains. If one takes a sliver of the broken lengths and squeezes it in the cross direction, a noticeably larger force is required to break it. The part is thus very strong in the direction of the chains and weak in the direction across the chains (refer to Chapter 2–chemical bond strength vs. secondary bond strength). This being the case, the strength of the part is not homogeneous and may result in a failure at a much lower than the anticipated force.
  - The orientation effect can indeed be used to one's advantage if the requirement is to have a predictably higher strength in a certain direction. Two common examples are polypropylene shopping bags and living hinges. In both cases the chains are purposely oriented in one direction. As shown in Figure 5.38, the hinge can have an almost indefinite fatigue life because of the controlled orientation. In the case of the bag, the handles are very strong in the length direction and take next to no force to tear in the cross direction.
  - When subjected to elevated or even room temperature, the chains will reorient themselves into their natural state, causing warpage and dimensional change. Figure 5.37 shows what happens after adding boiling water to the cup. The cup on the left has the original shape. The one on the right takes a bulge in the middle. The molecular chains have simply stress-relieved and reoriented themselves. *Caution! This can also happen during use to a part that had a shrink fixture applied to it after being ejected from the mold to prevent it from warping!*

Figure 5.36 Effect of mold run too cold–orientation

Figure 5.37 Cold cup showing stress-relieving after adding boiling water to it

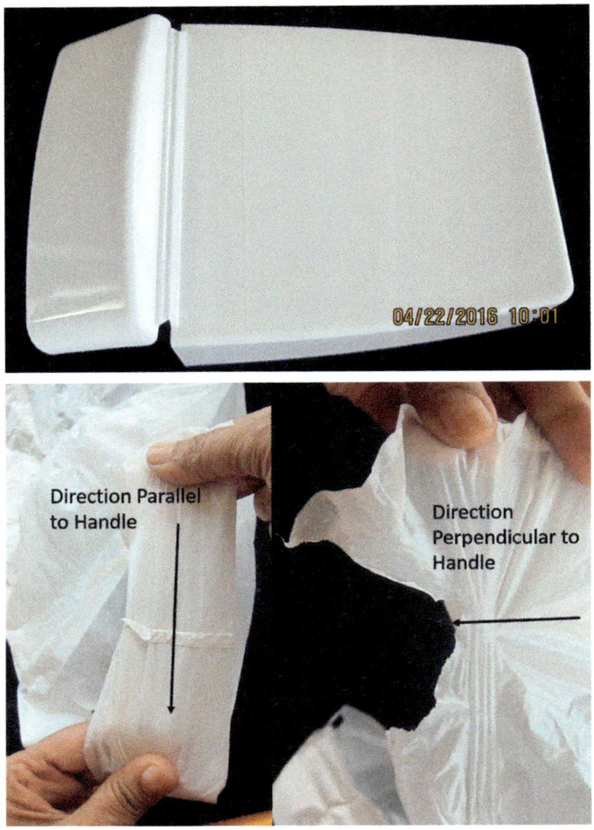

Figure 5.38 Taking advantage of orientation—living hinge and polypropylene bags

- Inadequate crystallization in semicrystalline plastics. These materials need a certain temperature and time in order to develop the orientation. Parts cooled inadequately will:
  - Be brittle
  - Continue to develop orientation as they age, resulting in warpage and dimensional changes

5.5.8 Gate Blush

There are many factors that can contribute to gate blush [8]; however, the underlying cause is erosion of the frozen wall around the gate area during the fill stage of molding.

As illustrated in Figure 5.39, the plastic melt enters the mold via an orifice or gate. Once inside the mold cavity, the hot melt in direct contact with the cooler mold solidifies and ceases to flow. Pressure from the injection unit forces additional molten plastic into the cavity and develops a fountain flow profile as described in Chapter 3, Section 3.2.6.1.2.1.

While the cavity is being filled, the growth in thickness of the solidified layer is controlled by local heat transfer, which is controlled by local shear heating near the wall and heat transfer into the cooler mold steel. Ideally, during fill there is a balance between shear heating and heat transfer to the mold, resulting in a solidified layer or skin that is relatively uniform in thickness.

Very often, the solidified layer opposite the gate is eroded by additional very hot material that feeds into the mold as the component is being filled out. This erosion is caused by locally higher shear forces developed by locally higher melt velocities, resulting in the external surface blush. As mentioned in Chapter 3, this reduced or eroded area not only manifests itself as a visual defect, it also reduces the level of protective compressive stress on the outer surface, making that area weaker in impact and more susceptible to environmental stress cracking.

While there are a number of other "cosmetic" issues that can develop near a gate, such as record grooving and streaking, the most common form of gate issue is the blush due to erosion as described in Figure 5.39.

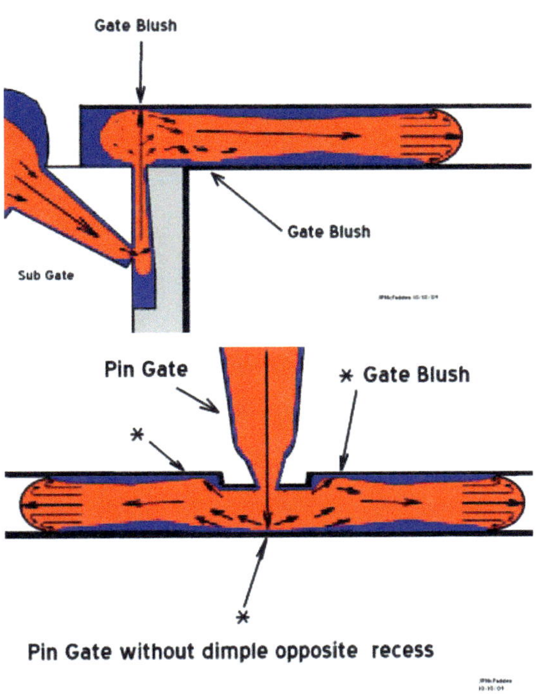

Figure 5.39 Gate erosion process [7]

One of the common methods to reduce the gate blush is by what is known as velocity profiling. The injection velocity is reduced in the first small fraction of the total injection phase to help reduce the shear during the wall formation in the gate area, to improve the thickness of the gate layer. Once this stable area is formed, it is less likely to erode during the rest of the injection phase.

5.5.9 Flow Marks, Gloss, and Texture Variation

There are multiple reasons for flow marks. Some of these are:

- Local burning or degradation of the plastic due to high shear
- Pigment or other additives separation from the resin or degradation due to poor compatibility of the pigment or additives or overheating. See Figure 5.40.

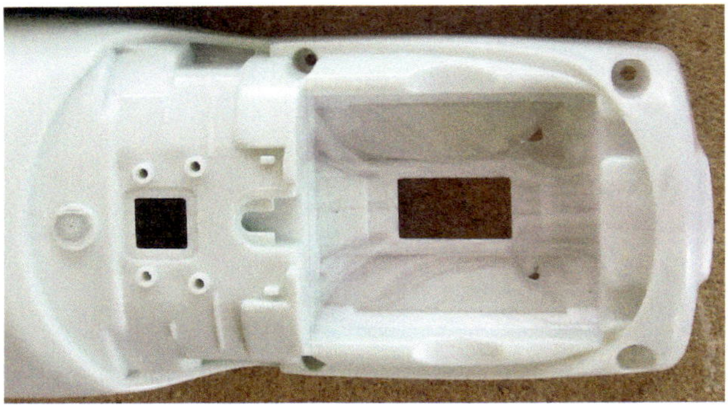

Figure 5.40 Flow marks due to poor compatibility of the pigment

- Inadequate homogenization of the material because of a large shot size relative to the barrel
- Degradation of the material because of a small shot size relative to the barrel
- Too much or too little back pressure
- Contaminants in the material including incompatible resins (such as nylon in PC)
- Glass fibers on the surface in glass-filled resins
- Splay due to excessive moisture or too long a residence time; see Figure 5.41

Figure 5.41 Splay due to moisture in the resin [9]

- Trapped air or gases
- Incomplete melding of the flow fronts
- Hesitation
- Sudden change in direction of flow
- Sudden change in the thickness
- Ribs being too thick
- Sharp internal corners
- Non-uniform packing resulting in shiny and matt areas near each other
- Non-uniform heating or cooling of the mold including hot spots; see Figure 5.42
- Non-uniform heating and/or packing causing the texture to appear more or less matt; see Figure 5.42

Figure 5.42 Noticeable change in the reproduction and matt level in the textures with the change of mold temperature and/or pack pressure

- Non-uniform texture or polishing of the mold walls
- Too high or too low graphics interrupting the flow of the material. See Figure 5.43 top. One way to minimize the knit line appearance is to provide venting for the trapped air and gases to escape. In the middle of a part (as opposed to the edges, where conventional venting can be employed), this can be achieved by inserting a laminated section of steel right under the graphics with controlled venting gaps between them. See the bottom image in Figure 5.43.

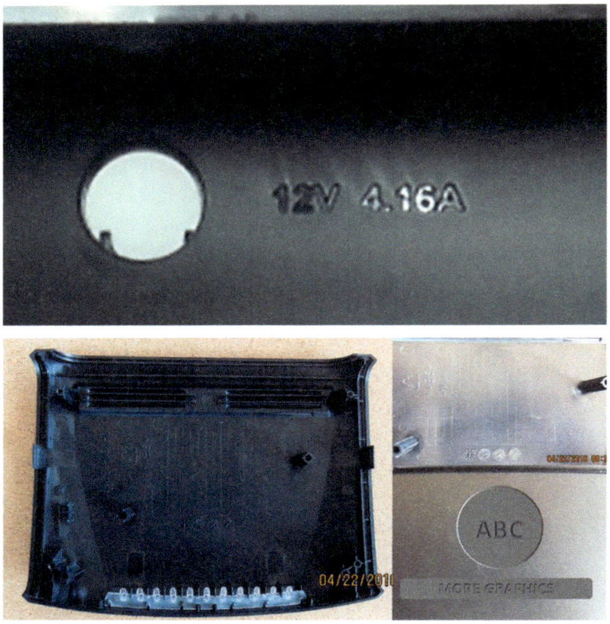

Figure 5.43 Deep graphics causing interruption of flow

Because of the fact that flow marks are caused by multiple reasons, there is not one standard method to eliminate or reduce them. Instead, a careful and logical cause and effect study should be conducted with tools like a fishbone diagram, as will be explained in Chapter 8.

5.5.10 Record Grooving

Record grooving are concentric grooves that look like the old fashioned vinyl records–hence the name. These occur as the flow front hesitates, builds up pressure again, flows, and hesitates again. This is the result of a less than adequate pressure at the flow front and of inadequate velocity and/or stock temperature. Adjusting the pressure or velocity, gate size, or location will generally improve or eliminate this. See Figure 5.44.

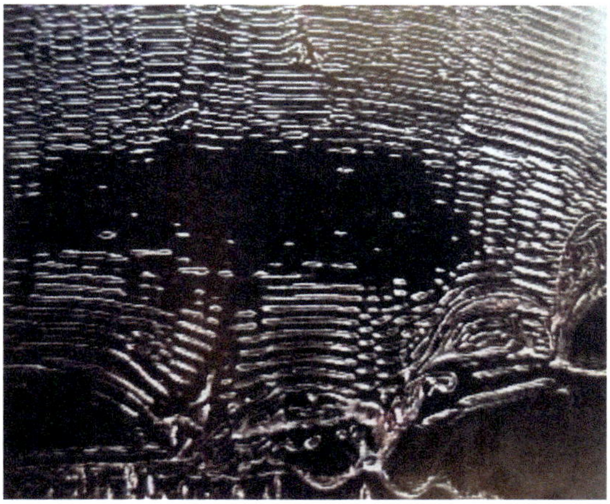

Figure 5.44 Record grooving in part [10]

5.5.11 Warping

Warping at the highest level is caused by:

- Different shrinkage within adjoining areas of the part, or
- Distortion due to poor ejection

Let us look at each one separately.

5.5.11.1 Non-Uniform Shrinkage

This can be caused by:

- Sharp variation in thickness (thin sections shrink less than thicker sections)
- Different shrinkage in different directions in semicrystalline materials or glass-filled materials. See Figure 5.45. The two discs are produced in the same mold. The left one is molded in nylon with different shrinkages in flow and cross flow directions (0.012-0.018) and the right one is molded in PC with uniform shrinkage in both directions (0.005-0.007). In the case of the nylon, the peripheral direction shrinks a lot more than the radial direction, causing it to "oil can."
- Non-uniform packing (poor placement of gates, inadequate packing pressure)
- Non-uniform heating or cooling; see Chapter 3, Figure 3.16
- Higher than acceptable mold temperature difference between the core and the cavity side

Figure 5.45 Warpage (oil canning) due to different shrinkage in flow and cross flow directions

5.5.11.2 Poor Ejection

If a part is ejected with an inadequate number of—or unbalanced—ejector pins, it is distorted during ejection while still very hot, resulting in permanent warpage. Also, if the part is difficult to eject due to an inadequate draft, the part will be distorted in the process of ejection, causing the same effect.

5.5.12 Flashing and Short Shots

Flashing can be due to:

- Inadequate clamp tonnage
- Too high an injection speed and/or overfilling in the injection phase
- Overpacking
- Damaged parting line
- Vent gap too high for the material being used
- Unbalanced cavities

Short shots can be caused by:

- Inadequate packing pressure
- Low mold or melt temperature
- Low injection speed and/or underfilling in the injection phase
- Inadequate gate size
- Undersized gates and/or inadequate number of gates and or/ improper location of the gate(s)
- Higher than acceptable length to thickness ratio in the part
- Unbalanced cavities

5.5.13 Sink Marks

The main reason for sink marks is a local cross-sectional area that is more than 25 percent larger than the nominal wall, in the case of amorphous materials, or 15 percent, in the case of semicrystalline materials. The causes for short shots given above can also contribute to the underfilling of the parts.

Flow simulation can detect areas susceptible to sink marks that are otherwise difficult to detect in a complicated shape. Looking at Figure 5.46, the areas in red have a much higher volumetric shrinkage than the adjoining areas, predicting likely sink marks.

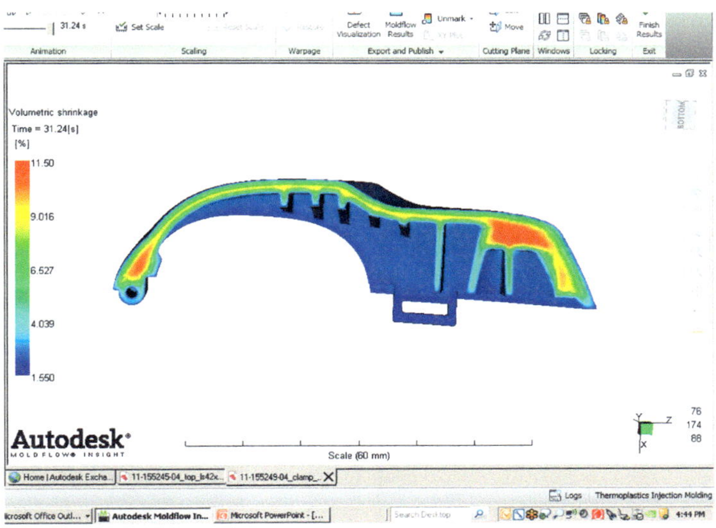

Figure 5.46 Sink mark prediction in Moldflow [8]

5.5.14 Scuffing

Scuffing is mostly caused by inadequate draft—especially in textured areas. It can also be caused by uneven ejection due to an inadequate number of pins, or unbalanced ejection causing the part to rotate slightly in the process of ejection, thus dragging one edge.

Additionally, it can be caused by local protrusions in the mold walls due to poor polishing.

5.5.15 Read Through and Shadows

Read through and shadows can have the same reasons as sink marks when the local cross section is only marginally smaller than the maximum allowable. Additionally, sharp internal corners and abrupt variation in the wall thickness will cause read through on the opposite wall. This also applies to internal embossed or debossed characters or symbols on the inside wall.

See Figure 5.47. A software tool called DFMPro [11] correctly predicted the flaws in this part. More information about DFMPro will be provided in Chapter 9.

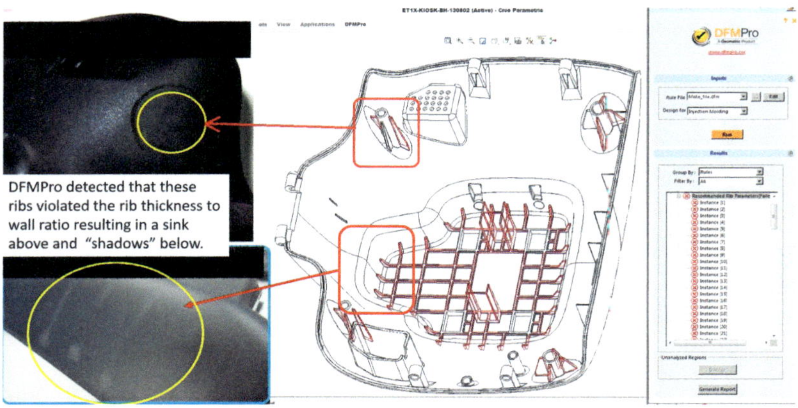

Figure 5.47 The figure shows unacceptable read through in the thick and thin areas

5.5.16 Unequal Shrinkage in Cross Direction

Please see the discussion on warpage. As discussed before, semicrystalline and glass fiber materials have substantially different shrinkage in the flow and cross flow directions. Figure 5.45 shows this effect in the same part molded in nylon and PC. The periphery is in the cross flow direction from the centrally gated part. In the case of PC, this has no effect, since the shrinkage is isotropic. However, in the case of nylon, the higher shrinkage in the periphery causes it to shrink considerably more, causing the oil canning effect. This may also happen in the case of a glass-fiber-filled part.

5.5.17 Use of Shrink Fixture

It is not uncommon to see wooden and/or metal fixtures being used after the part has been ejected, to reduce the warpage or control shrinkage to bring a dimension into specification. In my view, and also in the view of many of the experts in the field, this is only a short term solution for a few reasons:

- The part will go back to its original shape when exposed to a temperature in the proximity of its heat distortion temperature (HDT), such as when being placed next to a sunny window
- Even at room temperature lower than its HDT, the part will continuously tend to creep towards its original shape–especially in the case of lower HDT materials such as polystyrene, polyethylene, and PVC

An even more important issue is the induced stress that is introduced in the process of keeping the part from shrinking. This will reduce the mechanical performance and chemical resistance of the part.

See Figure 5.37 and the accompanying discussion for a clear PS cold cup going back to a more natural shape when subjected to elevated temperature.

5.6 Scientific Molding

It is important to now have a quick discussion on scientific molding, which, as the name indicates, basically adds a more scientific approach to the traditional molding. Before we go into the details, let us quickly look at the difference between traditional and scientific molding.

5.6.1 Traditional Molding vs. Scientific Molding

Injection molding evolved from a manual process in the 1940s, when machines were virtually unsophisticated arbor presses in which plastic was squeezed into a mold that was manually clamped by hand [12]. Material was fed into a cylinder, heated by heater bands, and a plunger was used to squeeze the melted plastic into a cold mold. Pressure was maintained until the part solidified, and then the mold was opened and the part taken away. The main controls were pressure and time, dictating how hard and how long the plastic was squeezed into the mold.

As the process became more sophisticated, a pump was added to the machine and the plastic was injected using hydraulic force. Early machines still had only one stage–squeeze. The pressure was set so that the plastic filled the mold; the part was packed to an appropriate level and held there until solidified. If a part was insufficiently filled, it was squeezed more. If a part was overfilled, it was squeezed less.

As the machine evolved further, two-stage machines became standard. But the technique remained the same—squeeze the plastic into the mold and pack it during the first stage, and then simply switch to a smaller pump during the second stage (or holding phase), primarily to conserve energy. The holding pressure was left either at the same pressure as in the first stage or, in some cases, slightly lower to minimize overpacking at the gate.

This was largely the technique of choice over the next 20 years and it is still widely used throughout the world. However, during the 1970s, there was an increase in the understanding of rheology, which made clear that some advantages could be gained by using an alternate molding methodology called Decoupled Molding[1]. Before we move on to scientific molding, let us review two molding-related phenomena:

As discussed before, in non-Newtonian fluids, the viscosity goes down as the shear rate or the velocity of injection is increased. In traditional molding, the velocity of the squeezed material was virtually uncontrolled because of the poor understanding of the phenomenon.

The speed of fill is limited by the lack of the required pressure available in the machine, creating a "pressure-limited" condition. The first two stages (filling and packing) are coupled together and cannot be controlled independently. Even worse, if only one pressure setting is used throughout the cycle, all three stages (filling, packing, and holding) are coupled together.

In "Decoupled Molding," the mold is filled at a controlled velocity until the cavity is approximately 95% full. At this point, the pressure is reduced to a lower holding pressure and is used to pack the mold. Because of the sensitivity of the process to injection time and pressure, sophisticated machine repeatability and controls are required.

5.6.2 Details of the Process

The filling of the cavity can be divided into injection, pack, and hold [13].

In the injection phase, the mold cavity is almost completely filled with the softened plastic. The plastic begins to cool down as soon as it hits the walls of the mold and forms an outer frozen skin. As the injection continues, the plastic flows in between the frozen layers. The reason why the cavity is not completely filled in this phase is

[1] Decoupled Molding is a service mark of RJG Inc.

to prevent flashing of the part, at the very low viscosity and high pressure during this phase, with the slightest variation of the injection speed. It also makes sure that a cushion is created during every shot. See earlier discussion on the topic.

As the plastic cools, the molecules start to get closer to each other, resulting in shrinkage. At this point, if the addition of plastic were to be stopped, the parts would be underpacked, showing visible sink marks and ripples. Figure 5.48 shows a fully packed and an underpacked part.

Figure 5.48 Fully packed part (left) and underpacked (right)

To avoid this and to compensate for the volumetric shrinkage, the injection phase is followed by the pack phase (generally combined with the hold phase), where the amount of plastic equivalent to the volumetric shrinkage of plastic is packed into the mold. At the end of this phase, the amount of plastic in the cavity must be equal to the theoretical weight of the part, which is equal to the solid density of the plastic multiplied by the volume of the cavity. Less plastic will result in an underpacked part and more in an overpacked part, leading to higher than necessary molded-in stress and/or flashing of the part at the parting lines.

If the pressure on the screw is removed right after the part has been packed, the plastic may flow back into the barrel, as the gate may still be unfrozen. A balanced pressure that prevents flashing of the part while preventing backflow, pack, and hold pressure is applied until the gate is frozen. In some cases, the holding pressure may need to be profiled to produce acceptable parts. Profiling stands for controlled and stepped lowering of the pressure at various intervals within the phase.

Figure 5.49 shows these phases. Underpacked parts result in defects such as sinks and voids and usually exhibit post-molding shrinkage, leading to dimensional instability. Overpacked parts may have frozen stresses that get relieved after molding, resulting in defects like warpage or lower mechanical and chemical performance.

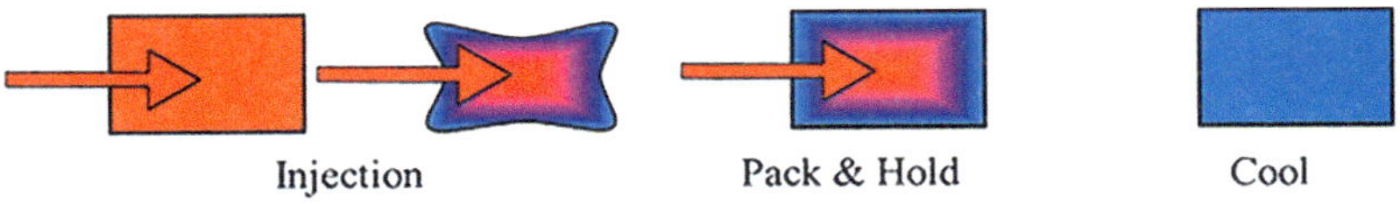

Figure 5.49 Injection molding phases [13]

5.6.3 Establishing the Optimum Process

5.6.3.1 Step 1: The Viscosity Curve and the Injection Speed

As discussed before, as the shear or injection speed is increased, the flow changes from a non-Newtonian one to one approaching Newtonian. The random chains become aligned parallel to each other and the crystalline areas open up. See Figure 5.50.

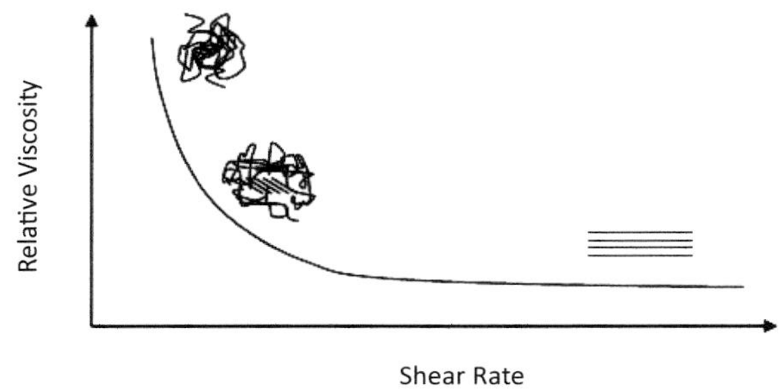

Figure 5.50 Opening of the crystalline areas with increase of shear [13]

The material is subjected to a large amount of shear forces during the cavity filling stage. The shear rate is proportional to the injection speed. If the shear rates are in the non-Newtonian region of the curve, small variations in the rate will cause a large shift in the viscosity. This will make the mold filling inconsistent, resulting in shot to shot inconsistency. It is therefore important to find the Newtonian region of the curve and set the injection speed (and, therefore, the shear rate) within this region. The viscosity curve can be generated at the molding machine for any given mold. This is also known as in-mold rheology.

The effect of shear rate on viscosity is much greater than that of temperature. Therefore, as long as the actual melt temperature is within the recommended range, a similar profile will be developed.

It is important to know the intensification ratio of the screw from the machine manufacturer. See discussion in Section 5.1.2.3. The viscosity (Y axis) is determined using the following formula:

Viscosity = Peak Injection Pressure × Fill Time × Screw Intensification Ratio

Figure 5.51 shows a typical viscosity curve generated at the molding machine.

Looking at the curve in Figure 5.51, one can notice that the viscosity stays fairly constant after about 60 % of the injection speed. Therefore, setting the injection speed to 70 % would ensure that the filling stage of the process will stay consistent. Any small natural variations will not cause large changes in viscosities, resulting in repeatable shot to shot quality of parts. This is especially important in the case of tight tolerance parts and multicavity molds. Optimizing the injection speed through in-mold rheology is only the first step to achieving a robust process and it is followed by the holding phase and the cooling phase, as described a little later in this chapter.

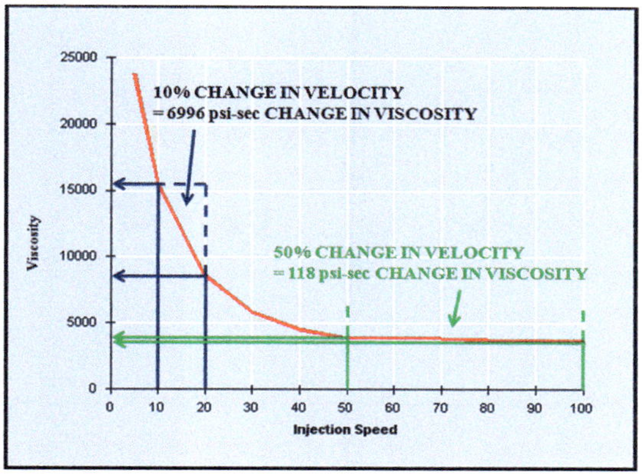

Figure 5.51 Typical viscosity curve [13]

5.6.3.2 Step 2: Balancing the Cavities

As the plastic flows through the runners into the cavity, the melt has a given temperature, pressure, and velocity. All three variables are time dependent and will change within the short time the plastic flows, until it reaches the end of the fill.

Now consider a two cavity mold with identical parts being molded in each cone. If the two cavities do not fill under exactly the same conditions, based on the above discussion, the parts produced from each cavity will be different. To produce identical parts it is imperative that the flow in the two cavities be balanced, ensuring that the plastic has reached the end of the fill under the same conditions in both of them.

A systematic study needs to be conducted to balance the cavities. The main reason for the lack of balance may be (along with the suggested corrections):

- Major difference between the cavities–thickness, size, flow length. In this case, processing and/or tooling correction may not help.
- Uneven heating of the cavities. Readjust or repair cooling lines.
- Leakage in one of the cavities. Repair parting line.
- Unbalanced or obstructed runners. Balance runners.
- Poorly designed or placed gates. Optimize gate size and locations.
  - Note: already optimized gates should not be changed to address unequal flow.
- Poor or blocked vents. Correct and repair vents.

A series of progressive short shots are produced for this study. The best way to measure the degree of fill is to compare the weights at various points. See Figure 5.52 for a typical graph from a four cavity mold study.

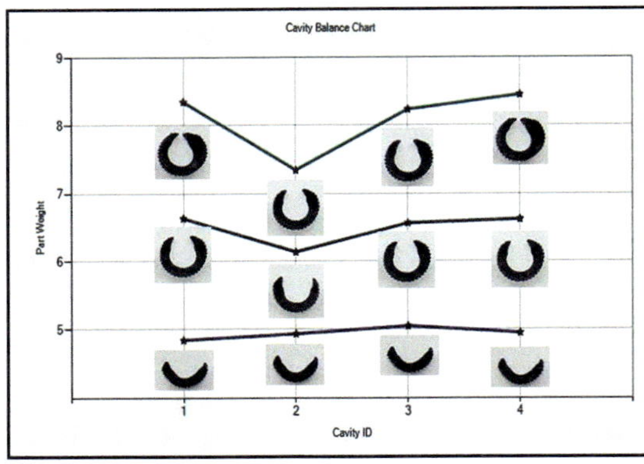

Figure 5.52 Cavity filling at various stages of progression of the fill [13]

5.6.3.3 Step 3: Pressure Drop Study

It is important to minimize the pressure drop from the nozzle to the gate of the part so that adequate pressure is available to fill the part without overburdening the molding machine. Unnecessary pressure drops can be attributed to the nozzle diameter and length, sprue and secondary sprue diameters and lengths, runner layout, length, shape, and size, and obstructions in any of them.

Figure 5.53 shows the results from a typical study. Delta P refers to the difference between the peak pressure needed and the maximum pressure available from the machine. It is recommended that the Delta P be at least 10% higher than the peak pressure.

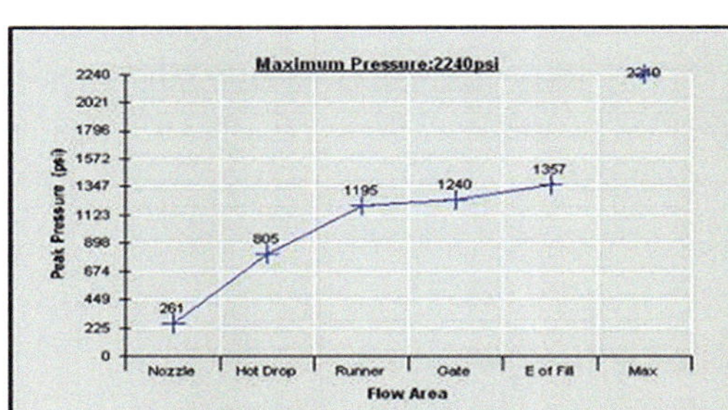

Flow Area	Peak Pressure	% of Maximum	Delta P	% Delta P
Nozzle	261	11.7	261	11.7
Hot Drop	805	35.9	544	24.3
Runner	1195	53.3	390	17.4
Gate	1240	55.4	45	2.0
E of Fill	1357	60.6	117	5.2
Max	2240	100.0	883	39.4

Figure 5.53 Pressure drop from the nozzle to the end of fill [13]

Looking at Figure 5.53, if the maximum available hydraulic pressure is 2,200 psi, the end of fill pressure should be less than 2,200 psi. If it is not, the screw needs more pressure to move at the set injection speed than the one available from the machine. This condition is called being "pressure limited."

A capable molder will look at the pressure drop and look for areas that are causing the pressure to be high. The cause may be as simple as a blocked runner or may be more complicated and require simulations to optimize the path from the nozzle to the gate.

5.6.3.4 Step 4: Setting the Pack and Hold Pressures

As discussed before, this phase packs the cavity with an amount of plastic equivalent to the volumetric shrinkage that occurs during the cooling of the plastic as it hits the cold walls of the mold. The variables are pack pressure, hold pressure, pack time, and hold time. In most cases, packing and holding are the same and are collectively called the holding phase.

The ideal holding pressure is determined by evaluating the process window of the mold. The process window is also called the molding area diagram. This is the area in which acceptable parts are molded. The bigger the window, the more robust is the process.

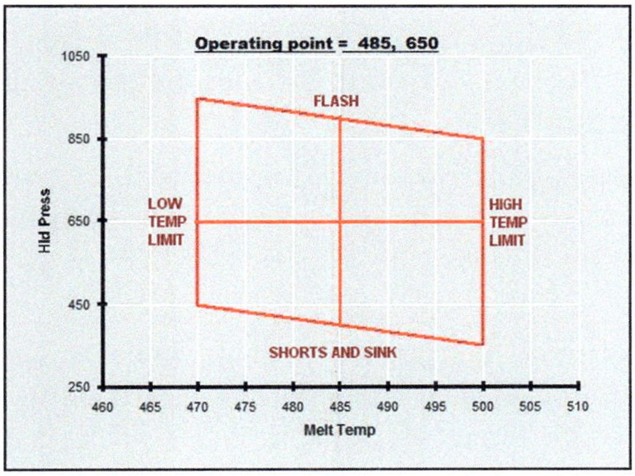

Figure 5.54 A typical molding window [13]

Figure 5.54 shows a typical molding window. Any point inside the window will produce an acceptable part in terms of visual, dimensional, and functional requirements specified for the part. Outside the process window, the parts can either have sinks or flash below or above the holding pressure limits. Outside the melt temperature limits, there may be unmelted plastic below the lower limit or degraded plastic over the limit. These are both detrimental to the part quality. The process should be set as close as possible to the center to get the most robustness. As stated before, the larger the window, the more robust is the process.

In the case of semicrystalline materials, because of the defined melting point, the allowable temperature range is very small. In this case, the window study is conducted at a fixed melt temperature; the variables in this case being the holding pressure and the mold temperature. The actual mold temperatures must be measured and the water temperature set to maintain this. The mold temperature is critical, since it provides the energy for the molten molecules to reach their final equilibrium resting states with minimal molded-in stresses. Mold temperatures are especially critical for crystalline materials, to provide the energy for the crystalline areas to go back to their natural bundled state. If the mold temperature and time within the mold does not allow for most of the crystallization to occur within the mold, the crystallization process will continue at a slower place outside the mold, leading to lower mechanical and chemical performance, and possibly severe warpage and unpredictable change in dimensions.

5.6.3.5 Step 5: Setting the Hold Time

As long as the gate is not frozen, the plastic can enter or leave the cavity. The holding pressure must therefore be applied until the gate is frozen. Failure to do this may cause the material to leak back into the nozzle. (As a matter of fact, this is done purposely in some parts like large trash cans with the gate at the bottom, to reduce the stress at the gate to prevent premature failure.) However, in most cases it will cause the part to be underpacked and/or introduce an unnecessary part to part variation, reducing the process robustness.

A very simple study is performed to determine the time it takes for the gate to freeze. Samples molded with different holding times are weighed and the time after which the weight remains constant indicates that the gate is frozen. This is set as the hold time.

See Figure 5.55. It shows that the part weight remains constant after 9 seconds. The holding time is set to about one second longer than the gate seal time, to account for possible shot to shot variation.

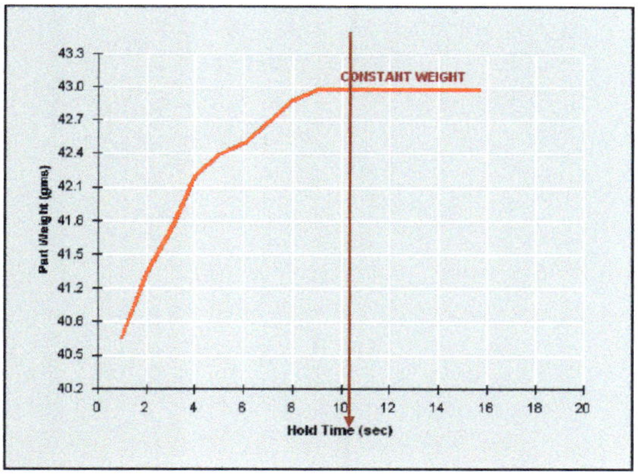

Figure 5.55 Gate freeze study: this shows the gate being frozen at about 9 seconds and the hold time set to about a second more [13]

5.6.3.6 Step 6: Setting the Cooling Time

Although the plastic starts to cool down as soon as it hits the walls of the mold, an additional time is required at the end of hold. The part needs to reach a temperature where it becomes stiff enough to be ejected without being distorted. The time required to reach this state is called the cooling time, and it should also be set so that the part dimensions remain unchanged beyond a controlled amount outside the mold. Although a longer cooling time may not harm the part, it will certainly be a profit killer.

The determination of the optimum cooling time can get complicated. For parts with thick sections, it is difficult to measure the internal temperature at the center of the thickest section and there is an uneven heat transfer within the mold. It may also take many hours for the mold temperature to stabilize within the mold. Shrinkage can also be influenced by changes in cooling times.

Figure 5.56 shows that some dimensions may be more sensitive than others. Dimension A is not influenced by the cooling time range within the study. However, dimension B changes with the cooling time. The target value for dimension B is 0.135 inches. To compensate for this, one should either set the cooling time at around 17 seconds or make changes to the steel to reduce the cooling time while achieving the same dimensions. Identifying the lower and upper limits will also provide a graphical representation of where the cooling time can be set.

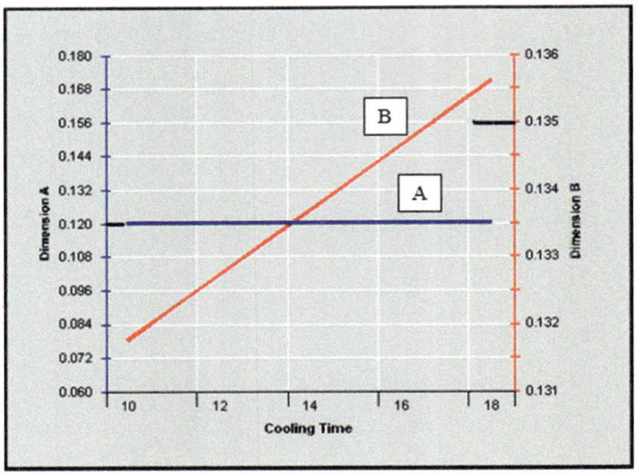

Figure 5.56 Graphical representation of the required cooling time [13]

■ 5.7 Putting It All Together—Processing Parameters Recording

Two of the major weaknesses of most molders today are proper recording and control of the process.

5.7.1 Proper Recording of the Process

As discussed throughout the book, the processing of a part can affect the visual, dimensional, performance characteristics, and the cost of a part. The following is a typical approval process for a part:

- Preliminary visual and dimensional inspection of the part by the responsible engineer and corrections as needed
- Form, fit, and functional tests including critical tests on the ruggedness of the parts
- Engineering changes to meet any deficiencies

There may be many iterations of the above until all the requirements are met.

The next step is a full FAI (First Article Inspection) of the part including dimensional, visual, and form and fit by the quality control department. The FAI report from the molder is also accompanied by a CoC (Certificate of Compliance) to assure that all that is specified in the documentation of the part, including RoHS, UL, material, etc., is in compliance.

After a successful approval of the above by the engineering and manufacturing department, the part is released to production. Only limited representative characteristics need to be monitored for subsequent incoming lots.

What is missing in the above process, typically, is an exhaustive processing sheet certified by the manufacturer.

A processing sheet incomplete in its details is similar to a FAI with certain critical dimensions not measured. Also, any change in the processing from the initial submission can not only change the dimensions, but also the performance of the part.

Let us now look at what needs to be included in the processing sheet. From the foregoing, it should now be very clear how these parameters can affect the performance of the part.

Temperatures:

- Mold
- Melt
- Water (including difference between inlet and outlet)
- Drying air temperature
- Dew point of drying air

Pressures:

- Injection (plastic—not just hydraulic)
- Hold (plastic—not just hydraulic)
- Back pressure (plastic—not just hydraulic)

Speeds and time:

- Fill time
- Hold time
- Mold open/close time
- Cooling time
- Recovery (time from the start of filling to its end)
- Gate seal time
- Total cycle time
- Material residence time
- Drying time

Volumes, weights, and distance:

- Barrel size
- Shot size
- Shot weight as percentage of barrel capacity
- Transfer point

- Cushion
- Water flow rate
- Reynolds number
- Hopper capacity

Material and part and general information:

- Setup date
- Setup person
- Resin type, grade, trade name, manufacturer
- Resin color
- Part number and revision

Mold and machine information:

- Machine number and make
- Mold number
- *Intensification ratio* (This is critical information to convert the hydraulic pressure to plastic pressure. The only way a part can be consistently molded on different machines is by having constant plastic pressures!)
- Screw diameter
- Maximum pressure available

A full Excel sheet that records all of the above is available at *www.vikpedia.org*.

5.7.2 Ongoing Control of the Process

In addition to accurately recording all of the above information, the manufacturer should ensure that no changes to the setup are allowed by the shop personnel. A proper procedure should be in place to make sure that changes are only made by qualified personnel, and that they are accurately recorded and discussed with the customer, if they are outside the allowable pre-established range.

Figure 5.57 shows an actual processing sheet for a highly engineered part. Figure 5.58 shows its English translation. The following information is missing or has errors:

Temperatures:

- Melt
- Water (including difference between inlet and outlet)
- Drying air temperature
- Dew point of drying air

Pressures:

- Injection (plastic)
- Hold (plastic)
- Back pressure (plastic)

Speeds and time:

- Fill time
- Hold time
- Mold open/close time
- Cooling time
- Recovery (time from the start of filling to its end)
- Gate seal time
- Material residence time
- Drying time

Volumes, weights, and distance:

- Barrel size
- Shot size
- Shot weight as percentage of barrel capacity
- Transfer point
- Cushion
- Water flow rate
- Reynolds number
- Hopper capacity

Material and part and general information:

- Resin type, grade, trade name, manufacturer (it just says PC)
- Resin color
- Part revision

Mold and machine information:

- Machine number and make
- Mold number
- Intensification ratio
- Screw diameter
- Maximum pressure available

In addition, observe the following "remarks" and my comments related to them:

1. "Due to variable in environment temperature, material flow and electric voltage, it can be adjusted within the tolerance range of this recorded molding parameter setting."

- Very general and subjective instructions

2. "In case of exceeded tolerance range due to different material or huge deviation from production, new molding parameter to be re-established, documented, audited, and to be approved."

- Very general and subjective instructions

3. "Temperature allowable tolerance: ± 10 °C"

- Arbitrary and too broad a range—especially for mold temperatures

4. "Pressure allowable tolerance: ± 8 kg.f"

- Arbitrary

5. "Speed allowable tolerance: ± 8 %"

- Arbitrary

6. "Screw position allowable tolerance: ± 5 mm"

- Arbitrary

7. "Time allowable tolerance: ± 5 seconds"

- This is really dangerous. The *total* injection time may be less than 5 seconds!

It is obvious that the sheet not only does not have all the desired information, it has WRONG information. It seems very likely that changes can be made at will by the shop personnel.

With these kinds of controls, will any two lots be similar in dimensions or performance?

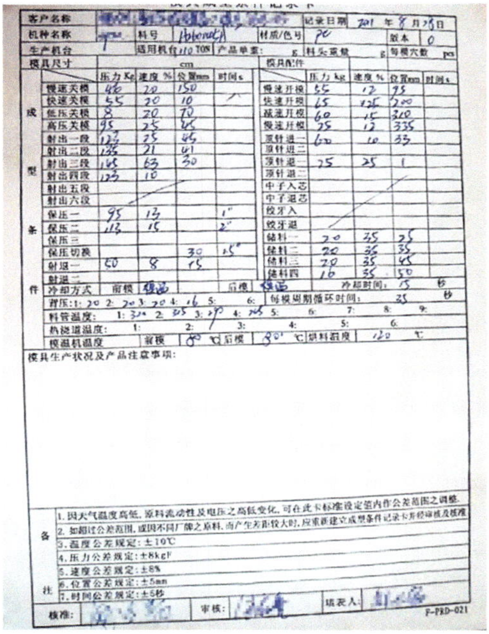

备注
1. 因天气温度高低，原料流动性及电压之高低变化，可在此卡标准设定值内作公差范围之调整。
2. 如超过公差范围，或因不同厂牌之原料，而产生差距较大时，应重新建立成型条件记录卡并经审核及核准
3. 温度公差规定：±10℃
4. 压力公差规定：±8kgF
5. 速度公差规定：±8%
6. 位置公差规定：±5mm
7. 时间公差规定：±5秒

核准： 审核： 填表人：

F-PRD-021

Figure 5.57 Example of poorly filled molding process recording sheet

Customer Name							Record Date	2001 Year 8 Month 8 Day		
Model Name			Part Number	XXXXXX		Material / Color	PC		Revision	0
Production Machine			Machined Used	110 TON	Part Weight	g	Runner Weight	g	Cavity per shot	pcs
Mold Size		cm				Mold Auxilliary				
		Pressure (kg)	Velocity (%)	Position (mm)	Time (sec)		Pressure (kg)	Velocity (%)	Position (mm)	Time (sec)
Molding Parameter	Low Speed Mold Close	46	20	150		Low Speed Mold Open	55	12	85	
	High Speed Mold Close	55	20	10		High Speed Mold Open	65	25	200	
	Low Pressure Mold Close	8	20	70		Slow Down Mold Open	60	15	310	
	High Pressure Mold Close	95	25	45		Low Speed Mold Open	25	12	335	
	Injection Stage 1	122	25	45		Ejection 1				
	Injection Stage 2	135	21	41		Ejection 2	75	25	1	
	Injection Stage 3	145	63	30		Ejection Retract 1				
	Injection Stage 4	123	10			Ejection Retract 2				
	Injection Stage 5					CORE In				
	Injection Stage 6					CORE Out				
	Holding Pressure Stage 1	95	13	1 inch		Screw Forward				
	Holding Pressure Stage 2	113	15	2		Screw Recovery				
	Holding Pressure Stage 3					Screw Suck Back 1	20	35	25	
	V/P Transfer Point		30	1.5 inch		Screw Suck Back 2	20	35	35	
	Nozzle Retract 1	50	8	15		Screw Suck Back 3	20	35	45	
	Nozzle Retract 2					Screw Suck Back 4	16	35	50	
	Cooling Method	CAV			CORE		Cooling Time			seconds
	Back Pressure	6:				Cycle Time 25				seconds
	Barrel Temperature	1:320	2:305	3:290	4:265	5:	6:	7:	8:	9:
	Hot Runner Temperature	1:	2:		3:	4:		5:		6:
	Mold Temperature Controller Temperature		CAV	80 °C	CORE	80 °C	Material Dryer		120	°C
Production Tool and Part condition that need to be alerted:										
Remarks	1. Due to variable in environment temperature, material flow and electric voltage,it can be adjusted within the tolerance range of this recorded molding parameter setting.									
	2. In case of exceeded tolerance range due to different material or huge deviation from production, new molding parameter to be re-established, documented, audited and to be approved.									
	3. Temperature allowable tolerance: ± 10 °C									
	4. Pressure allowable tolerance: ± 8 kg.f									
	5. Speed allowable tolerance: ± 8%									
	6. Screw position allowable tolerance: ± 5 mm									
	7. Time allowable tolerance: ± 5 seconds									
Approved by:				Checked by:			Recorded by:			

Figure 5.58 English translation of processing sheet in Figure 5.57 [14]

5.7.3 Allowable Changes to the Processing Parameters

So, what *are* the allowable changes to the processing parameters?

This is a frequently discussed question. The common answer may be something like ± 5% or even worse, from the example above. This is very arbitrary and can lead to a lot of unacceptable variations.

The correct answer is based on the process window study discussed earlier. Changes may be made only within the window. Even so, it is not a good idea to have a setting very close to the edge of the window, as it will reduce the overall robustness of the process.

5.8 References

[1] Adapted from *http://elitemachinerysystems.com/*

[2] Images provided courtesy of SABIC

[3] Adapted from John Bozzelli; Plastics Technology, December 2011

[4] Adapted from Autodesk; *https://knowledge.autodesk.com/support/moldflow-flex/learn-explore/caas/CloudHelp/cloudhelp/2016/ENU/MoldflowInsight360/files/GUID-167DA746-9E44-4F4D-96C6-0D0998C3EF97-htm.html*

[5] Adapted from *https://www.ozeninc.com/wp-content/uploads/2014/11/fig04B-3.jpg*

[6] Suhas Kulkarni; *Robust Process Development and Scientific Molding*, 2nd ed., Hanser Publishers, Munich (2017)

[7] Courtesy of Joseph P. McFadden Sr.; *http://McFaddenCAE.com*

[8] Adapted from article by Joseph P. McFadden Sr.; *http://McFaddenCAE.com*

[9] Suhas Kulkarni; *http://fimmtech.com*

[10] Ulf Bruder; *User's Guide to Plastics*, Hanser Publishers, Munich (2015) p. 204

[11] DFMPro is the trademark of Geometric, Ltd.

[12] Adapted from *http://www.plasticstoday.com/almanac-fundamentals-decoupled-molding/8741599362942*

[13] Adapted from articles by Suhas Kulkarni – FIMMTECH, a consulting firm that provides services in the area of injection molding of plastics. FIMMTECH is also in the process of developing products that will increase efficiency of the molding process, educate personnel, and better manage the molding facility. One of the first products to be released is the software "NAUTILUS," which helps in the development of robust and optimized processes. Suhas Kulkarni also teaches a course on injection molding at the University of California, San Diego that can be offered as in-house seminars. For more information, please visit *http://www.fimmtech.com*.

[14] Translation courtesy of Teng Weng Onn

[15] Courtesy of Covestro LLC

[16] Courtesy of Paul Koehn and Flexcraft

6 Secondary Operations

6.1 Introduction

There are few injection molded plastic parts or assemblies that do not have some kind of secondary operation applied to them. At the highest level, these can be divided into three broad categories, namely:

- Assembly or joining
- Machining or some other kind of material removal process
- Decorative, printing, painting, plating, etc.

Assembly or Joining

From the short and long term performance perspective, this is by far the area that needs special consideration in material selection, design, tooling, and processing. The main operations are:

- Ultrasonic welding
- Thermal staking
- Vibration welding
- Solvent bonding
- Insert molding
- Conventional fastening

Machining

These consist of conventional machining operations mainly to remove gates and runners and occasionally to create an opening in the part that is not economical or practical as part of the molding process.

Decorating, Printing, Painting, Plating, Vacuum Metalizing, Etc.

These operations involve adding graphics, paint, or metal coat to a part for both visual and functional reasons, such as electrical conductivity for electrostatic discharge. The main operations are:

- Pad stamping
- Silk screening
- Laser etching
- Water transfer decoration
- Painting
- Plating
- Vacuum metalizing

In-mold decoration can also be used for adding graphics to plastic parts. These can be both two- and three-dimensional.

Since the most common performance failures are related to ultrasonic and thermal assembly, the rest of the discussion in this chapter will be concentrated on these two operations. Numerous sites will be provided in the reference chapter (Chapter 9) for those needing more information in other than the abovementioned areas.

6.2 Ultrasonic Assembly

6.2.1 Theory

Before we go into the details of the process, let us discuss some basic principles that make ultrasonic welding and assembly possible.

6.2.1.1 Hysteresis Heating

When a metal strip is bent back and forth in a rapid fashion, the friction between the moving internal particles against each other results in heat. "Hysteresis," or the lagging of an effect behind its cause, thus takes place. The dissipation of the heat cannot keep up with its generation and the temperature at the point of flexure goes up. See Figure 6.1.

Figure 6.1 Bending a metal strip back and forth resulting in heat. Courtesy of Emerson [1]

Similarly, if a plastic part is made to be compressed or stretched and released in a rapid fashion, the generation of the heat at the point of compression and stretching raises its temperature in the area undergoing the change. In my seminars, I have the attendees stretch a rubber band back and forth rapidly and feel the heat generated against their upper lip. Try this at home!

In ultrasonic welding, a very small area is molded at the end surface of the part and it is compressed and released rapidly by an external force or *vibrated* at a very high frequency, causing it to be locally heated to the point of melting or flowing. This small area is called an "energy director." See Figure 6.2. More on energy directors starting in Section 6.2.6.1.1.

Figure 6.2 "Energy director" under rapid vibration resulting in melting and bonding. Courtesy of Emerson [1]

6.2.1.2 Energy for Ultrasonic Welding

Before we go into the actual mechanism for generating the required energy through vibration to melt the plastic, let us revisit some basic physics formulas.

Work, Power, and Energy

Work

Work is the result of a force acting on an object to make it move a certain distance. Thus,

Work = Force × distance moved

$W = F \times d$

In the metric system, this can be written in Newton × meter, or Joule (J).

Power

Power is defined as the rate of work. Or,

Power = Work / time

$P = W / t$

The above can be re-written as

$P = F \times (d/t)$

Or, if v is the velocity:

$P = F \times v$

Energy

Energy = $P \times t$

6.2.2 Generation of Vibration

In the basic form, the 50/60 Hz power supply is converted to 20 kHz (or 15 kHz or 30 kHz or 40 kHz as explained later) through a power supply. This is then supplied to a piezoelectric device that converts the electrical wave into a mechanical one at the same frequency. A booster then amplifies the waves. Finally, a custom-designed "horn" connected to the booster delivers the vibration at an optimum amplitude to the surface of the part being welded. See Figure 6.3.

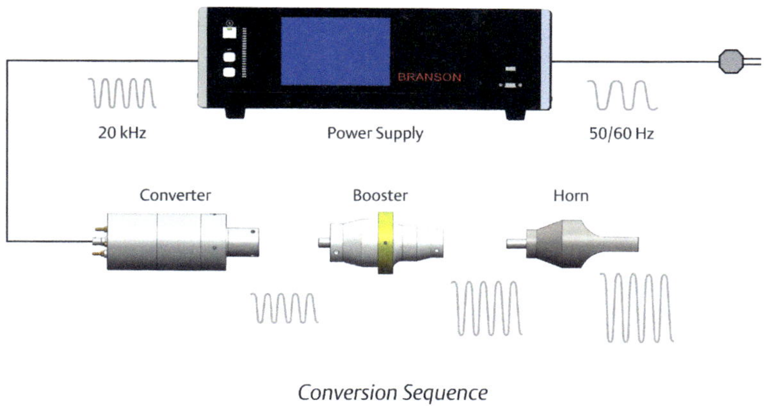

Figure 6.3 Generation of optimum vibration at the plastic surface. Courtesy of Emerson [1]

6.2.3 Ultrasonic Welding System Components

- Power supply
- Control level
- Actuator/stand
- Converter
- Booster
- Horn
- Fixture

Figure 6.4 shows typical welding machines. The pressure required on the horn is supplied from compressed shop air, and both the pressure and the down speed of the horn can be controlled by regulating the air supply. The rest of the controls can be contained in the frame shown (left) or be in a separate box on the side (right). The holding fixture is mounted on the table below the horn.

Figure 6.4 Typical ultrasonic machines. Courtesy of Emerson, showing the Branson Ultrasonic 2000 welder and the Branson Ultrasonic 2000X welder [1]

6.2.3.1 Control of Power and Total Energy Provided to the Welded Surface

As discussed before,

Energy = P × t

Looking at Figure 6.5, the power can be adjusted through the pressure and the down speed of the horn. The velocity can be controlled by the frequency or amplitude at the tip of the horn.

Finally, the total energy can be controlled by the time the vibration stays on.

In addition, another parameter needs to be controlled: the dwell time. This is the time the pressure stays on the weld after the vibrations have been turned off, to ensure that the joint has cooled down sufficiently not to spring back when the pressure is removed.

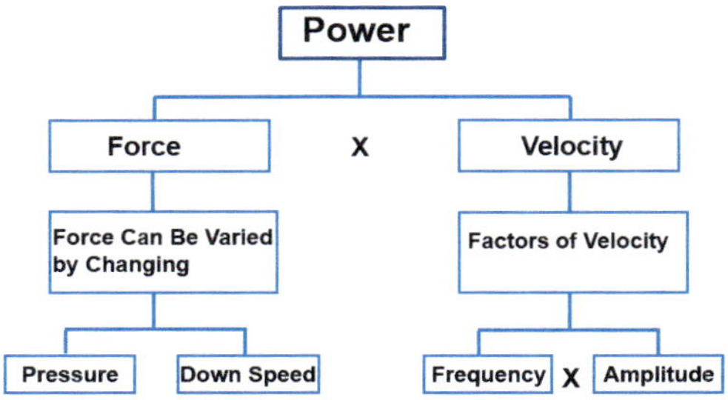

Figure 6.5 Power development in ultrasonic welding. Courtesy of Emerson [1]

6.2.3.2 Details of the Subcomponents

The assembly that receives the electrical signal from the power supply and delivers it as a controlled vibration at the tip of the horn is termed the "stack." It has the following components:

- Converter
- Booster
- Horn

Let us look at these in detail.

6.2.3.2.1 Converter

The electrical waves at the required frequency are converted into mechanical vibrations by the converter, also called transducer. The following description of its components and functions is adapted from the Dukane Guide to Ultrasonic Assembly [2].

The heart of a transducer is the piezoelectric ceramic elements, which, when exposed to electrical energy of alternating polarity, expand and contract dimensionally. As the electrical polarity changes from positive to negative, the ceramic elements get thicker and thinner. Although this dimensional expansion and contraction is small, its force capability is great.

As shown in Figure 6.6, the transducer driver assembly consists of the back slug, ceramic elements, and the front slug. Transducers use a steel back slug and an aluminum front slug. Since the acoustical transmission properties of aluminum are far superior to those of steel, this arrangement directs as much vibrational energy to the booster and horn as possible. The amplitude of vibration at the output end of most 20 kHz transducers is eight tenths of one thousandth of an inch (0.0008 inches or 20 microns).

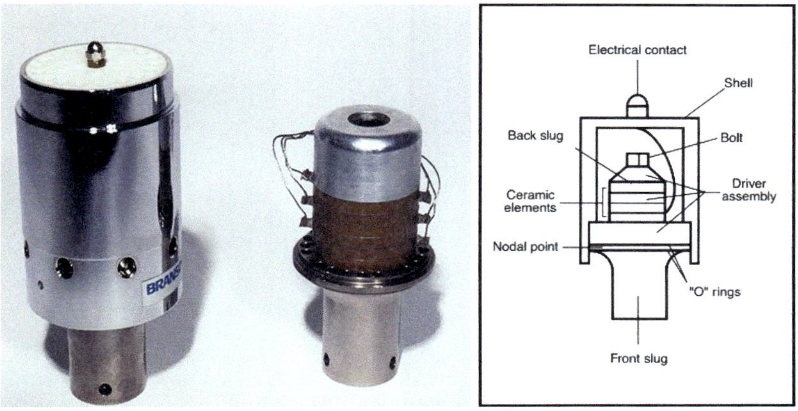

Figure 6.6 Converter assembly (Courtesy of Emerson) [1] and schematic [2]

There is an area near the center of the transducer where little or no motion occurs. This area is called the nodal point. Although there is little linear motion at the nodal point, there is radial expansion and contraction, so it is an area under great stress. The lack of movement makes the nodal point useful for mounting the transducer in a shell. Neoprene “O” rings insulate the shell from any residual vibrations. Anti-rotational pins are installed through the shell into the nodal point.

6.2.3.2.2 Booster

The next part of the stack is the booster. It is used to increase or decrease the amplitude to the one optimum for the assembly being performed, as will be described later in this chapter. This characteristic is described as the gain of the booster. This is dependent on the ratio of the input and output masses about a nodal point that does not vibrate.

There is a ring around the nodal point, which is used to fasten the stack to the machine frame. Looking at the booster fourth from the left in Figure 6.7, it can be seen that the mass on the input side of the nodal ring is twice the one on the output side. The gain is therefore 2. The first booster from the left reduces the amplitude with a gain of 0.6. The one second from the left has equal masses on both sides and therefore has no gain.

In the case of Branson, these are color coded and are available in aluminum (left) or stainless steel (right). The stainless steel ones have longer life. See Figure 6.7.

Figure 6.7 Boosters from Branson. Courtesy of Emerson [1]

6.2.3.2.3 Horn

The final component in the stack is the horn. See Figure 6.8.

Depending on the specific needs, the horns can be fabricated out of titanium, steel, or aluminum. Titanium has the best acoustical properties and high fatigue strength. Titanium horns are generally more expensive than horns made out of other alloys because of the higher cost of material and machining. Aluminum heat-treated alloys also have very good acoustical properties, and are used for lower amplitude, strength, or hardness. They can also be hard anodized or chromium or nickel plated for higher wear.

Figure 6.8 Various types of horns. Courtesy of Emerson [1]

For high wear applications such as welding glass-filled materials, hardened steel can be used. Because they tend to be brittle, they are generally used for low amplitude applications.

There are four main considerations in the design of horns:

1. Maximum amplitude at the tip

This is the most important factor in the design of the horn. The length of the horn must be optimized for mechanical resonance at the free end based on the ultrasonic frequency being used. This translates to one or multiple half wavelengths of sound in the horn material and configuration.

In order to have the maximum resonance and therefore the maximum amplitude, the horn has to be half wavelength of the frequency. For example:

Speed of sound in aluminum = 5,000 m/s (range: 3100-6400 m/s, depending on the alloy)

Wavelength at 20 kHz = 5,000 m/s / 20,000 Hz = 0.25 m

Half wavelength = 0.125 m or ~ 5 inches

2. Transfer of the vibration to the work surface without distortion so that there is uniform contact across it

Figure 6.9 shows the highly magnified distortion of the welding surfaces through modal analysis. If the horn is not adjusted for this distortion, uneven welding may result.

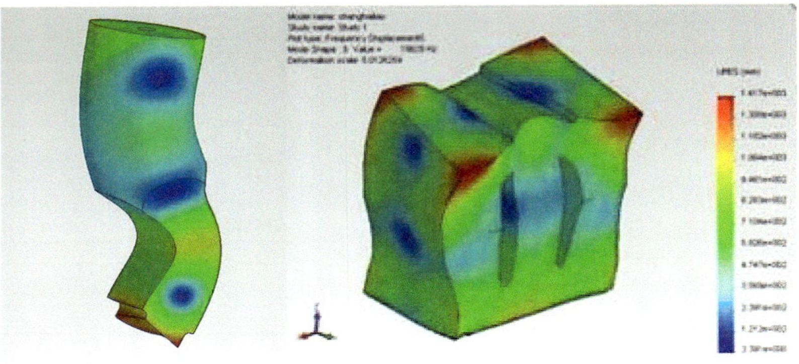

Figure 6.9 Modal analysis of horns show distortion of the horn body and therefore the welding surface [3]

3. Minimize stress in the horn for maximum life

Figure 6.8 shows various slots and hollow areas in the horn to archive this.

4. Optimum amplitude at the tip

This basically supplements the gain in the booster. As in the case of the booster, this can be accomplished by different masses in the top and bottom of the horn.

Figure 6.10 shows a stack that has a gain of 1 in the booster and 2 in the horn for a total gain of 2. For a 20 kHz system with an initial amplitude of 0.0008 inches or 20 microns, the amplitude at the tip would be about 0.0016 inches or 40 microns.

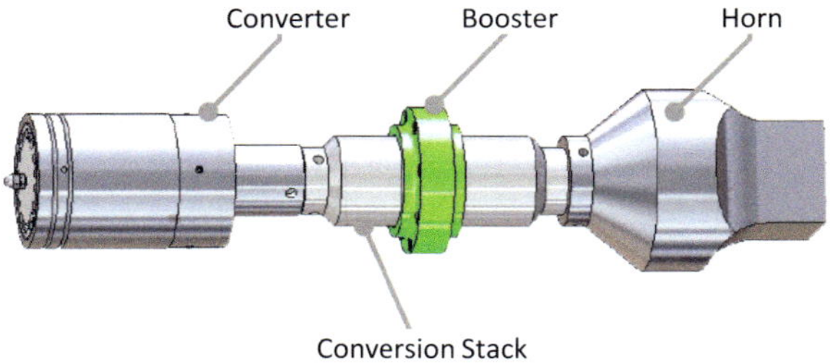

Figure 6.10 A typical stack. Courtesy of Emerson [1]

6.2.3.2.4 Fixtures

The final element in the ultrasonic welding system is the fixture, which reliably and consistently holds the part during the welding operation. Depending on the lifetime and specific requirements, fixtures can be made from epoxies to hardened steel.

The requirements for a good fixture are:

- Proper and level support of the part to be welded
- Stay clear of the horn surface
- Easy adjustment
- Stability
- Easy removal of the part
- Prevent scuffing or marking on the part

Figure 6.11 shows a typical fixture.

Figure 6.11 Typical holding fixture [2]

6.2.4 Types of Ultrasonic Welding

While 20 kHz ultrasonic machines are the most common, there are generally three other frequencies that these machines are made for:

- 15 kHz
- 30 kHz
- 40 kHz

Since the wavelengths are inversely proportional to the frequency, the total height of the stack and therefore the size of the machine are the smallest for the 40 kHz and the largest for the 15 kHz.

The analogy I use for the areas where these machines are used is that of hammers used by a goldsmith and a blacksmith. For fine, delicate work such as jewelry, the hammer is very light and the strokes are very gentle. For heavy iron work, a sledge hammer, along with heavy strokes, is needed. Since the amplitude and force of the 15 kHz vibrations are much larger than the ones of the 40 kHz, for small, cosmetic, delicate parts a 40 kHz machine may be suitable. For large and non-cosmetic parts, a 15 kHz machine may be more suitable.

6.2.5 Welding Polymers

Going back to Chapter 2, the main types of polymers are:

- Thermosets
- Thermoplastics

As stated in that chapter, thermosets are irreversible in terms of chemical reactions and therefore cannot be made to melt or flow by supplying thermal energy. Therefore, they are not suitable for this process.

The two types of thermoplastics are:

- Amorphous
- Semicrystallline

Let us discuss them in terms of ultrasonic welding.

Figure 6.12 Amorphous and semicrystalline material structures

From basic physics, the specific heat is the amount of calories needed to raise the temperature of 1 g of material by 1 °C. Thus, for water the specific heat is 1 cal/g/°C, as it requires one calorie to raise the temperature of one gram of water by 1 °C.

The latent heat is the heat required to change the state from solid to liquid, or liquid to vapor, and the heat needed to be removed for the reverse changes. Thus, the latent heat of water for fusion (ice to liquid) is 80 calories per gram and the one for vaporization is 533 calories per gram.

Now let us refer to Figure 6.13. For amorphous materials, the relationship between the temperature rise and the heat supplied is linear from the Tg to temperatures where the plastic starts to flow. This is also the case for semicrystalline materials until the melting point–the temperature where the crystalline areas start to open–is approached. There is a heat of fusion (an equivalent of the latent heat) required at this point to create the melt. In the case of nylon this is about 31 calories per gram [5].

In terms of ultrasonic welding, amorphous materials such as polycarbonate allow a lot of latitude in exactly how much ultrasonic energy is required to effect the fusion of the two pieces of plastics being bonded. However, in the case of semicrystalline materials such as nylon, the energy has to be accurately controlled. Not enough energy, and the pieces will not bond; too much, and the overmelting will cause the joint to be overwelded, causing both dimensional and visual issues.

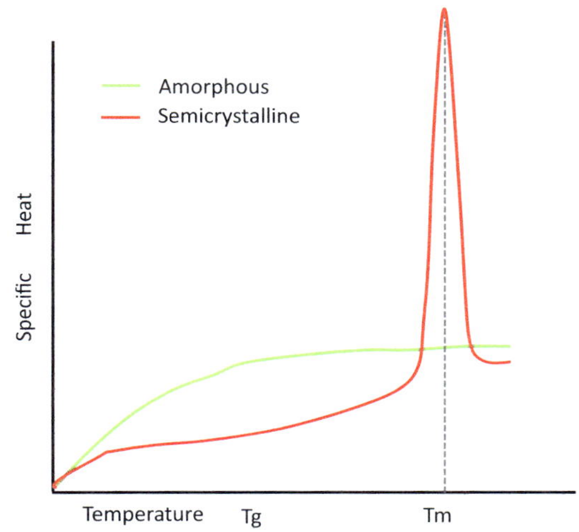

Figure 6.13 Tg, Tm, and specific heat of thermoplastics

Before we look at the ease of welding for various pairs of plastics, let us discuss near field and far field welding. See Figure 6.14.

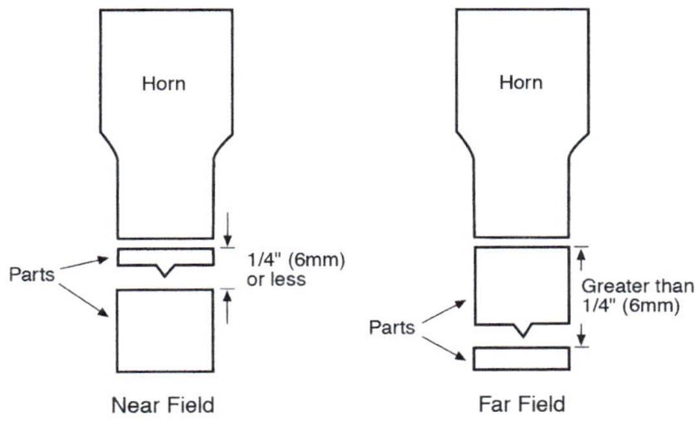

Figure 6.14 Near field and far field welding [2]

Near field and far field welding refers to the distance through which ultrasonic energy is transmitted, from the point of horn contact to the joint interface. When the distance between the horn and joint interface is 1/4 inch (6 mm) or less, it is considered near field. When the distance is greater than 1/4 inch (6 mm), the weld is considered far field. Whenever possible, it is best to weld near field. Far field welding requires higher than normal amplitudes, longer weld times, and higher air pressures to achieve a comparable near field weld [2].

Not all materials can be welded with equal ease. As stated earlier, amorphous materials are easier to weld than semicrystalline materials.

The softening temperature and modulus of elasticity also affect the weldability. A rigid material transfers the vibrations with relatively little loss to the energy director. Lower temperature materials need less energy. Thus, polystyrene, with its higher modulus and low flow temperature, is one the easiest to weld. Polycarbonate, with its lower modulus and higher temperature to flow, is more difficult.

6.2.5.1 Other Factors Affecting Ease of Welding

6.2.5.1.1 Hygroscopicity

The polymers listed below are hygroscopic and the moisture absorbed by them affects the welding. Polycarbonate parts, for example, left in a non-humidity controlled environment are more difficult to weld than those just molded.

- Nylon
- Polycarbonate
- Polysulfone

6.2.5.1.2 Drying Techniques

Poorly dried parts are more difficult to weld than those dried properly. This is mainly because the weld is weaker, as the chain size has been reduced.

6.2.5.1.3 Fillers, Filler Types, and Percentage

The following are some common fillers. Since these do not melt, they act as contaminants to the welding process.

- Glass
- Talc
- Calcium carbonate

In certain situations, since glass increases the modulus, it can help the efficient transmission of the ultrasonic energy to the joint. I have found that low quantity glass-filled nylon parts in far field welding performed better than the non-filled parts for this reason.

6.2.5.1.4 Other Contaminants

For the same reason as above, the following hinder the welding. The mold release additionally reduces the friction between the mating parts, thus retarding the frictional heat.

- Mold release
  - Silicone / Teflon
  - Paintable
  - Internal lubricants
- Colors
  - Pigments
  - Carriers
  - Amounts
- Flame retardants
  - Percentage
  - Type

6.2.5.1.5 Dissimilar Polymers and Other Factors

- Melt temperature
- Same polymer (different grades)
- Processing technology
- Melt index/flow rate
- Temperature

6.2.6 Design Factors

6.2.6.1 Joint Types

6.2.6.1.1 Butt Weld with Energy Directors

This is the most common design and it uses energy directors as mentioned before. Energy directors are required to concentrate the energy to a small controlled profile and volume of plastic. In the absence of one, the joint would need a much longer weld time and consequently energy to achieve a weld. In addition, the weld would

- Have poor part to part repeatability
- Have unacceptable height variation
- Have poor cosmetics due to the plastic oozing out from the sides

6.2.6.1.2 Energy Director Design

There is a lot of literature on proper design of energy directors from suppliers, such as Branson, Dukane, and others. Specific sites will be provided in the reference section (Chapter 9). See Figure 6.15 for guidelines for the most common design.

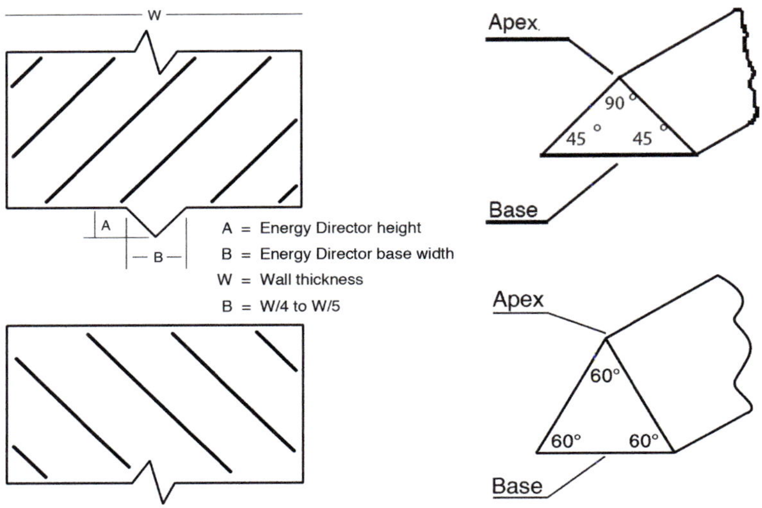

Figure 6.15 Energy director proportions for amorphous (top right) and semicrystalline material (bottom right) [2]

Properly designed energy directors:

- Increase joint strength
- Reduce flash
- Reduce weld time
- Require lower amplitude

Texturing the mating surface to the energy director increases the weld strength, reduces the energy requirement, and provides spaces for the "molten" plastic to flow into, thus avoiding an unsightly flash at the joint. These textures are typically 0.003 to 0.006 inches deep. More information on the texture in the reference section (Chapter 9).

6.2.6.1.3 Round Energy Directors

As seen in Figure 6.15, almost all energy directors historically have had triangular cross sections. The assumption is that the apex of this triangle is a sharp edge. However, due to molding and flow limitations, they seldom have this assumed sharp edge. In fact, the exact sharpness may vary from part to part and lot to lot. This fact results in a variation of the weld quality from one part to the next.

To overcome this challenge, Dukane Intelligent Solutions conducted extensive studies with round energy directors. The details of the round energy directors are shown in Figure 6.16.

Round energy directors, in conjunction with servo-driven actuators in place of the pneumatically driven ones in conventional machines, provide the highly precise control and repeatability that result in these joints providing considerably higher strength and repeatability than their traditional counterparts. The reference section (Chapter 9) has information on extensive literature on this new type of energy directors and the readers are encouraged to take the time to study them.

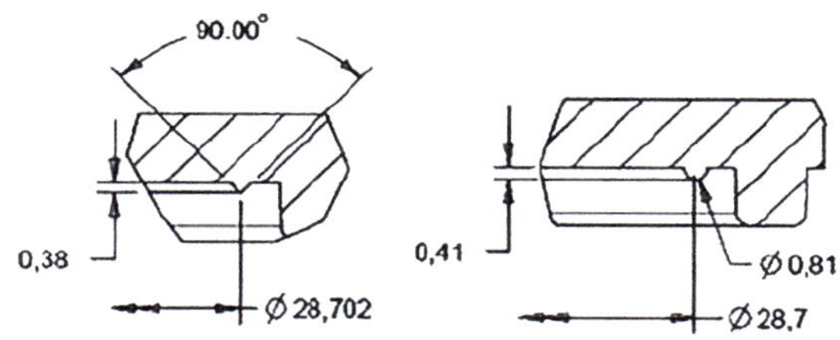

Figure 6.16 Conventional energy director (left) and round directors (right) [2]

6.2.6.1.4 Shear Joints

When the design allows, shear joints can have much greater strengths than butt joint. However, the design, processing, and fixtures needed for shear joints are a lot more complex and the ultrasonic equipment vendor should be consulted for the project to be successful.

Figure 6.17 shows a typical shear joint design.

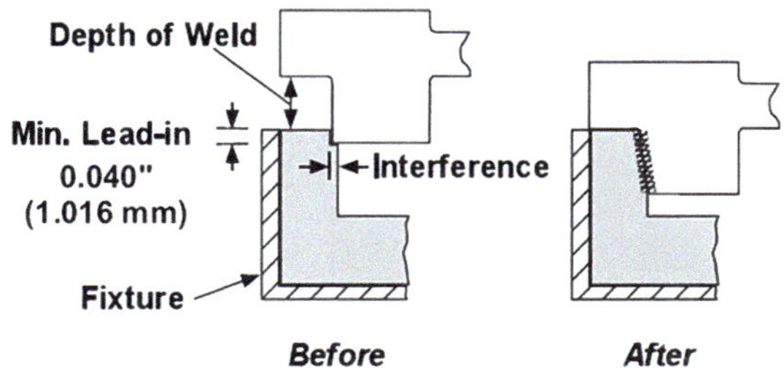

Figure 6.17 Shear joint before and after the welding operation. Courtesy of Emerson [1]

6.2.6.2 Part Design

As in molding, part design is an important consideration for ultrasonic welds, too. Let us look at some important considerations.

6.2.6.2.1 Sharp Corners

The discussion in Chapter 3 (reproduced below as Figure 6.18) considered a sharp corner near a weld being stressed back and forth 20,000 times or more in a typical weld cycle on a 20 kHz machine. Depending on the amplitude, this may fatigue the corner into a crack right at the machine or reduce the stress carrying capacity even if the crack does not occur during the welding cycle. It is therefore recommended that rules recommended for molding be also applied for the welding operation.

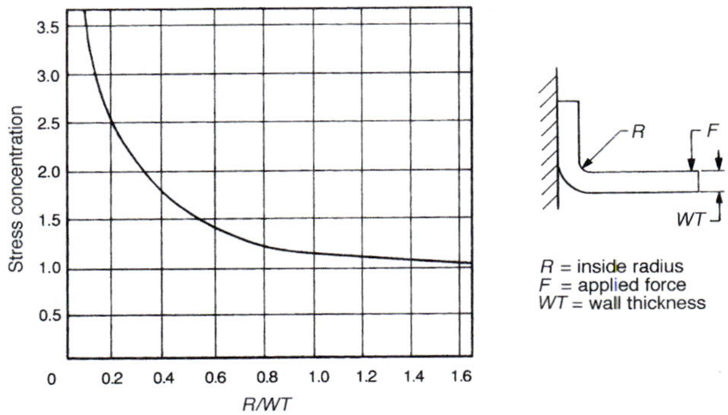

Figure 6.18 Relationship of inside corner radius to the stress concentration factor [6]

6.2.6.2.2 Holes and Bends

Any discontinuity in the plastic area transmitting the ultrasonic energy to the area being welded will cause this area to be underwelded in comparison to the area under a solid piece of plastic. To compensate for this many times, the energy will be increased to obtain proper welding in the deficient area. This in turn will cause the adjacent area to overweld, causing flashing and dimensional issues. Ultimately, the capability of the process will be substantially reduced.

6.2.6.2.3 Appendages

Appendages sticking out from near the weld may go into resonance, causing instant fatigue cracking or substantially reducing the fatigue life of the part.

6.2.6.2.4 Diaphragmming

A lid being welded to a box may vibrate wildly in the middle non-contact area, causing it to deform permanently or being marred.

6.2.6.2.5 Horn Contact

The horn should have a solid area under it to efficiently transfer the energy to the weld. A lack of this will cause weak and inconsistent welds.

6.2.6.2.6 Metal Contacts in Ultrasonic Assembly

It is not uncommon to sandwich metal contacts between parts being welded, as in electric connectors. The ends of these may go into uncontrolled resonance, causing the contacts to break off due to fatigue or having their fatigue life substantially reduced. They can also melt plastic surfaces they are in contact with, causing misalignment.

Figure 6.19 shows one such instance where contacts made out of beryllium copper were molded into a part (gray) that was subsequently welded to a mating part (yellow). The view on the top shows the alignment required for the proper position of the contacts with its mating part. The ribs in the top left view were supposed to provide a positive stop to the contacts for this purpose. After assembly, it was found that the contacts were severely misaligned. The cause was the ends of the contact going into resonance at the 20 kHz frequency, as shown by modal analysis (bottom right view), literally hammering the ribs at the back and melting them (bottom left view). The solution was creating an additional weld near and under the contacts and preventing resonance of the system as a whole.

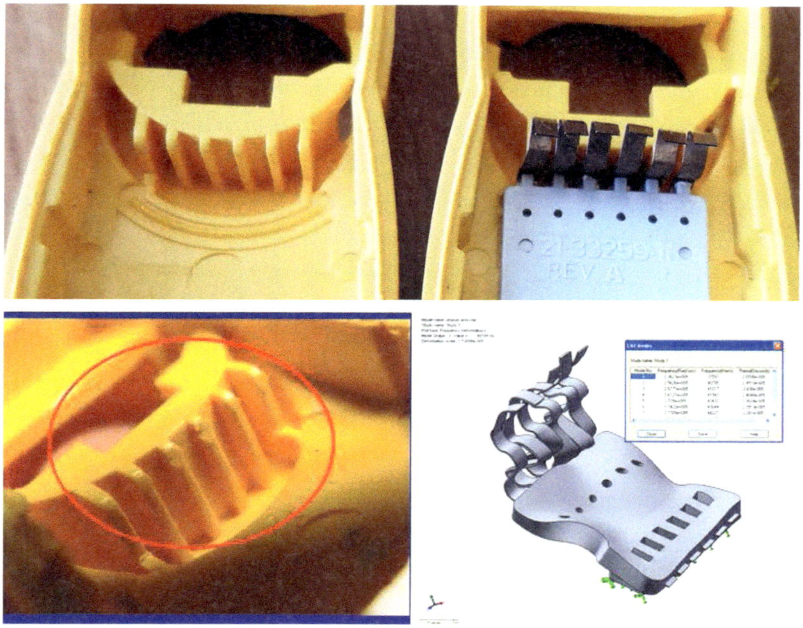

Figure 6.19 Metal contacts in assembly (top right) going into resonance (bottom right) and melting support area behind contacts (bottom left) [3]

6.2.6.3 Fixture Design

Proper fixture design complements and supplements the part design. The fixture has the following functions to perform for setting up and for welding cycles:

- Reliably provide a steady and level surface for the part to rest
- Reinforce the plastic part from lateral movement
- Prevent marring or scuffing of any surface
- Easy, fast, and repeatable placement and removal of the part without marring and scuffing any surface
- Provide clear area for the horn up and down movement
- Be easy to level

Fixtures can be made from epoxies, aluminum (hard anodized for longer life), or steel or any combination of the above. Epoxies are especially suitable for complicated and/or curved surfaces, as the cavity can be cast using the part itself as the pattern. The fact that they are inexpensive and fast to manufacture compensates for the fact that they do not last as much as their metal counterparts.

Figure 6.20 shows the two common types of fixtures. The left one is for conventional butt welding and the right for shear welding. Notice that in both cases the fixture is supporting both the bottom and left surfaces. This is especially needed for shear welding so the wall does not move to the left preventing a good, consistent weld. This prevents marking from the harsh metal edges on the metal. Another technique to prevent marring the top surface is to place a polyethylene film between the horn and the mating plastic surface.

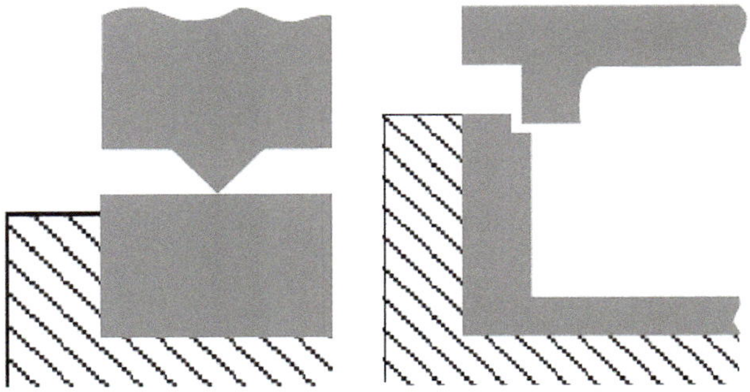

Figure 6.20 Proper fixturing for butt welding (left) and shear welding (right). Courtesy of Emerson [1]

6.2.6.4 Important Welding Parameters

The four wheels of a high performance automobile analogy can also be applied to the welding operation. So far we covered the materials; the stack, including the horn, and the fixture (tooling); and the part design, including the energy director. Now let us review the process.

The important welding parameters are:

- Amplitude
- Pressure
- Down speed
- Time
  - Weld
  - Hold

6.2.6.4.1 Amplitude

As before, the amplitude is defined as the peak to peak longitudinal displacement at the face of the horn (see Figure 6.21) and it has the most impact on the quality of the weld, as the heat generated at the joint interface is proportional to the square of the amplitude.

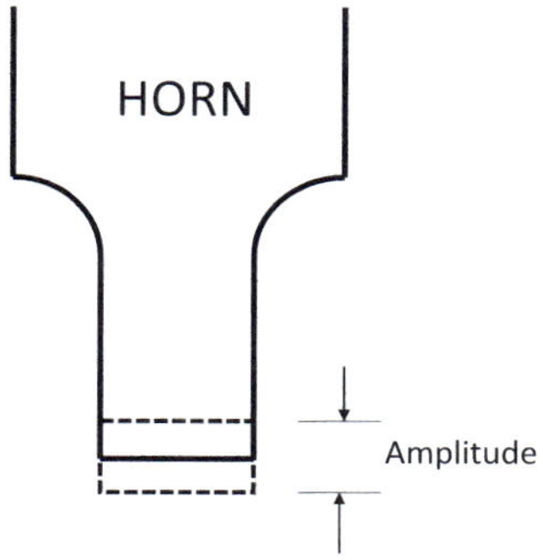

Figure 6.21 Amplitude

Table 6.1 shows the recommended amplitudes for some of the most common materials at various frequencies, in microns (μm).

Table 6.1 Suggested Amplitudes for Some Common Materials in Microns. Courtesy of Emerson [1]

Amorphous	15 kHz	20 kHz	30 kHz	40 kHz
Acrylonitrile Butadiene Styrene (ABS)	36-84	30-70	24-56	18-42
Acrylonitrile Styrene Acrylate (ASA)	36-84	30-70	24-56	18-42
Polycarbonate (PC)	72-120	60-100	48-80	36-60
PC/ABS	72-120	60-100	48-80	36-60
Polycarbonate/Polyester	60-120	50-100	40-80	30-60
Polyetherimide (PEI)	84-120	70-100	56-80	42-60
Polyethersulfone (PES)	84-120	70-100	56-80	42-60
Polymethyl Methacrylate (Acrylic, PMMA)	48-84	40-70	32-56	24-42
Polyphenylene Oxide (PPO)	60-108	50-90	40-72	30-54
Polystyrene (PS)	36-84	30-70	24-56	18-42
Polysulfone (PSU)	84-120	70-100	56-80	42-60
Polyvinyl Chloride (rigid PVC)	48-96	40-80	32-64	24-48
Styrene-Acrylonitrile (SAN)	36-84	30-70	24-56	18-42
Semi-Crystalline				
Cellulosics (CA, CAB, CAP)	72-120	60-100	48-80	36-60
Liquid Crystal Polymer (LCP)	84-144	70-120	56-96	42-72
Polyoxymethylene, Polyacetal (POM)	84-144	70-120	56-96	42-72
Polyamid (Nylon, PA)	84-144	70-120	56-96	42-72
Polybutylene Terephthalate (Polyester, PBT)	84-144	70-120	56-96	42-72
Polyethylene Terephthalate (Polyester, PET)	96-144	80-120	64-96	48-72
Polyetheretherketone (PEEK)	84-144	70-120	56-96	42-72
Polyethylene (PE)	108-144	90-120	72-96	54-72
Polyphenylene Sulfide (PPS)	96-144	80-120	64-96	48-72
Polypropylene (PP)	108-144	90-120	72-96	54-72

Amplitude Profiling

A more advanced technique is to "profile" the amplitude during the energy on phase, as shown in Figure 6.22. Starting with a higher amplitude, it is lowered to a smaller one within the phase. This results in:

- Increased weld strength
- Reduced part marking
- Reduced flash

It is recommended that the supplier have a machine that has the capability to profile the amplitude.

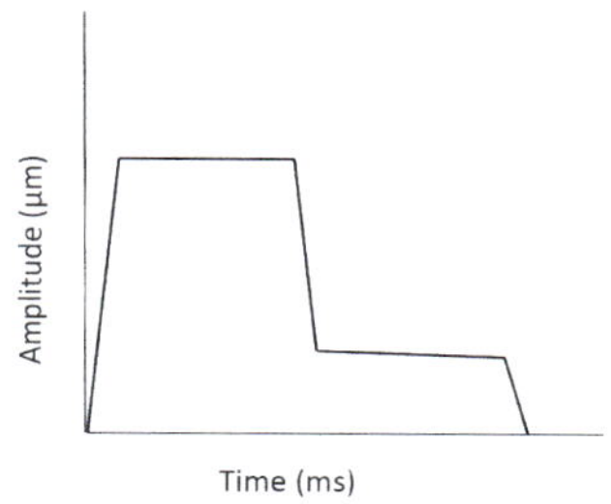

Figure 6.22 Amplitude profiling

6.2.6.4.2 Pressure

Adequate pressure ensures that sufficient force is developed to generate the required total energy. Additionally, the weld surfaces need to be kept under pressure to keep the joint from springing back and to promote the chain entanglement.

As a general guideline, the pressure should be maintained at around 20-30 psi for most welds. This needs to be adjusted to make sure excess pressure and down speed do not end up crushing the energy director instead of softening/melting it.

6.2.6.4.3 Time

Weld:

This is the time in which the weld surfaces are subjected to the set amplitude under constant pressure and down speed; this generates the heat that softens or melts the energy mating surface, and chain entanglement occurs between them.

Hold:

This is the time in which the welded surfaces are kept under constant pressure after the ultrasonic energy is turned off.

As a general rule of thumb, a weld time of one second and a hold time of two seconds is a good place to start.

6.2.6.4.4 Air and Voltage Regulation

Although modern machines have built-in voltage regulators, it is advisable to have regulated voltage input, especially in areas where there is a lot of voltage fluctuation.

It is also desirable to have a dedicated compressed air supply to the machine. A machine connected to the central air supply may lose cycle to cycle consistency if there is a sudden loss of pressure due to some other equipment drawing a lot of air. Also, if the air quantity is marginal, cycle to cycle consistency will be lost.

6.2.6.4.5 Monitoring

Modern machines can monitor consistent cycle to cycle quality through the following:

- Weld collapse: the distance the horn travels between the vibration start to the end of hold time; this ensures that the energy director is sufficiently collapsed
- Total energy: provides constant energy from cycle to cycle
- Distance: the horn travel stops at a set distance every cycle; this ensures a very controlled height for tight tolerance parts

6.2.6.4.6 Common Defects

The following are some of the common defects and their possible causes.

Marking:

This can be caused by sharp edges both on the horn and the fixture. It can also be caused by over or uneven welding. Improper leveling of the horn with the welding surface will make the horn dig in more on one side than the other.

Underweld, overweld, non-uniform weld:

The causes can be:

- Horn
  - Shape
  - Tuning
  - Damage
  - Connection

It is extremely important that the components within the stack are fastened to each other at the prescribed torque. If this is not the case, there will be a big uncontrolled energy loss resulting in many of the above defects.

- Improper holding fixture or fixture not aligned properly

See Figure 6.23. This is created by placing a carbon paper and paper between the horn and the part, and lowering the horn with a very small air or hand pressure. The impression on the paper will show a dark transfer of ink where the horn is touching the part surface, and a light or no impression where it is not. In the figure there is good contact at the top left corner and very little to no contact at the bottom right corner. This can be fixed by putting a shim at the bottom right corner of the fixture. Alignment plates are also available for very fine leveling of the horn.

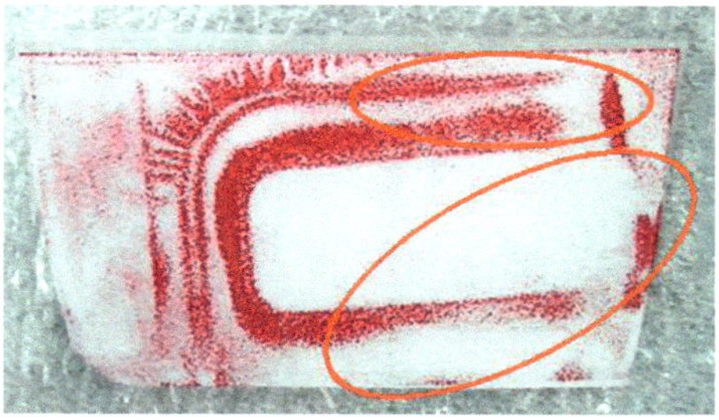

Figure 6.23 Checking for horn and fixture alignment

- Air pressure variation
- Voltage variation

Scuffing:

Mostly caused by higher than optimum amplitude. See Figure 6.24.

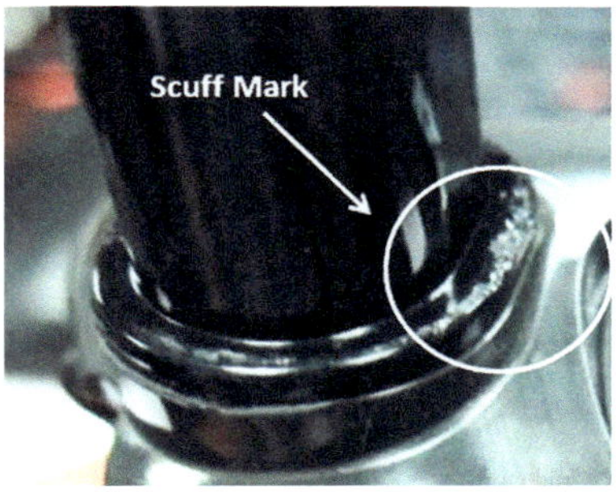

Figure 6.24 Scuff marks

6.2.6.4.7 Checking for Defects

See Figure 6.25.

- Underweld: On breaking the joint apart, the energy director appears bent, crushed, or deformed and may retain the original sheen. The properly welded surfaces generally have a whitening associated with a ductile failure.
- Non-uniform weld: The weld quality is non-uniform along the energy director.
- Overweld: The weld may extend beyond the energy director and may flash out of the joint.

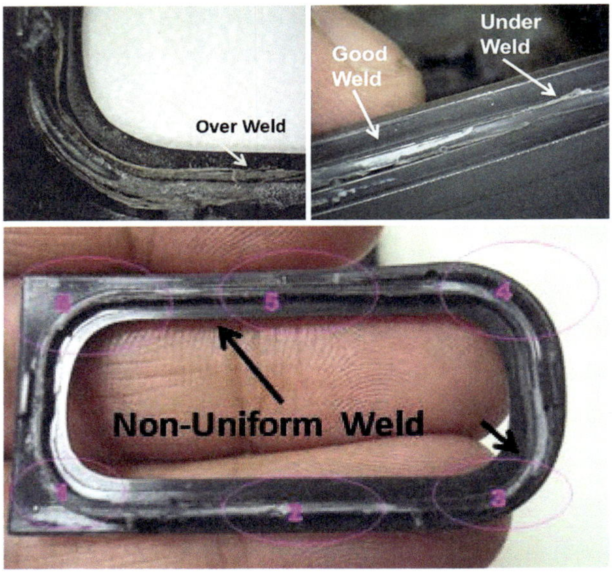

Figure 6.25 Examples of good, poor, and non-uniform welds

6.3 Heat Staking

Another common form of joining is heat staking. This process is used for:

- Heating in place threaded inserts within molded-in bosses to be used with the common machine screws for assembly. This is needed where the assembly will need to be fastened or unfastened multiple times, or to prevent an end user from using an improper screw to replace a screw designed for plastic that was fastened directly to the boss.
- Softening and reforming plastic standoffs to act as rivets to permanently join two pieces of plastic.

Both of the above can also be achieved ultrasonically. However, heat staking allows simultaneous staking of multiple locations even at different planes, which is not practical for ultrasonic welding because of the limited area of the horn. Additionally, it is not possible to insert or form at different levels because of the exact length needed by the horns, as discussed before.

The basic sequence of operation for both inserting and forming is:

1. Contact the plastic to be softened by a heated tip
2. Dwell until the plastic has sufficiently softened

3. Continue downward motion to the designated depth
4. Let plastic cool
5. Start upward motion

Although this appears to be very straightforward and simple, catastrophic failures can occur if all of the above steps are not followed, especially in the case of inserts. The main reasons for the failures are:

- Underheating of the parts and forcing the plastic to deform in a relatively cold state; this will cause long term hoop stresses in the boss, or cause the plastic to spring back in the case of forming
- Overheating of the boss and causing misalignment of the insert, causing the plastic to be under continuous stress when the mating screw tries to realign it
- Overheating the area being formed and thus reducing the cross section of the area under stress
- Overmelting, causing visual issues

See Figure 6.26. The top left view shows two plastic parts being joined by the post on the bottom part formed into a rivet; the heated tip on the top has the hollow shape that the standoff will be formed into. The top right view shows a properly formed rivet head after the heated tip is withdrawn. The bottom right view shows what happens when the tip travels too far: unsightly lumps of plastic ooze into unwanted areas; also, the cross-sectional area of the rivet is reduced. The bottom left view shows what happens if the horn does not travel far enough: again, the cross-sectional area of the rivet is reduced; in addition, there is a gap between the rivet head and the mating plate, resulting in a loose assembly. Note that the top piece does not have to be a plastic part; it can be a metal plate or a printed circuit board.

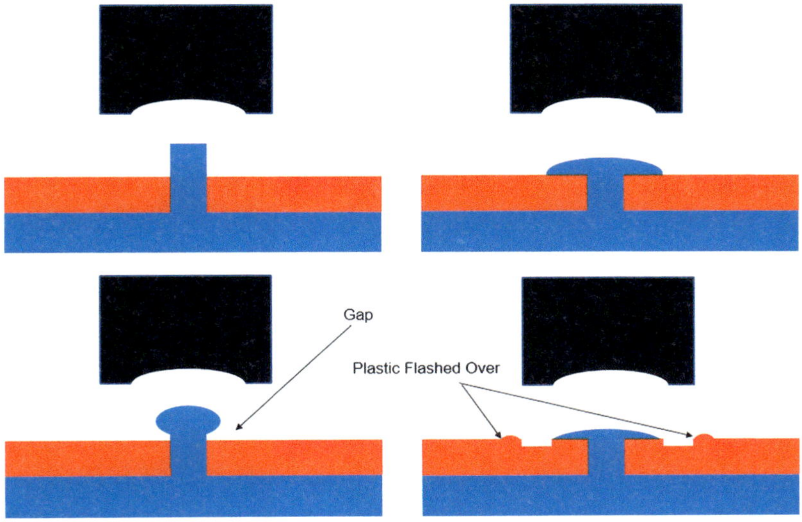

Figure 6.26 Properly and improperly heat staked parts

6.3.1 Heat Staking Machines

Figure 6.27 shows a properly configured heat staking machine with two tips. Figure 6.28 shows a heating tip with a heating band around it.

A properly configured machine has the following features:

- Precise air pressure control
- Precise down motion control
- Precise time and distance control for dwell and final stop
- *Individually* heated and controlled tips for each forming or inserting operation

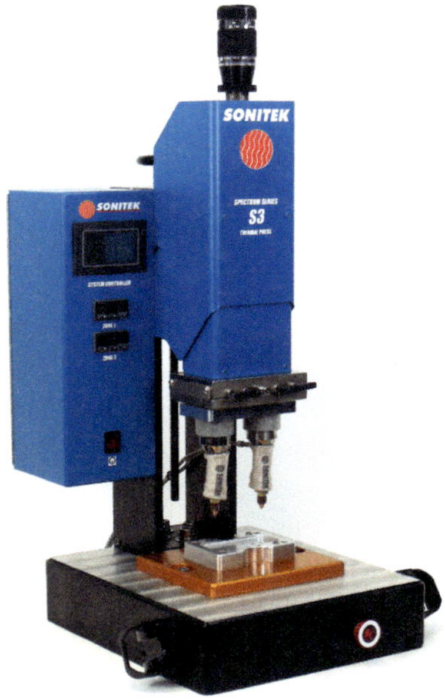

Figure 6.27 A properly configured heat staking machine and heated tip detail [4]

Figure 6.28 A properly designed heating tip and insert

Unfortunately, a lot of suppliers do not have the above controls in place. The machines they use have one common heater heating all the tips with one thermocouple. This results in vastly varying temperatures between tips, resulting in overmelting and undermelting. They also have one nonstop downward motion without a dwell. This results in the inserts being forced into their respective openings, causing high hoop stresses and the resultant catastrophic failures.

Figure 6.29 (top) shows a deficient setup as described above; Figure 6.29 (bottom) shows a properly configured system with individual heaters and controllers. Figure 6.30 shows actual failures due to the poor setup.

Figure 6.29 A poorly designed heat staking machine (top) and a properly designed one (bottom [4])

Figure 6.30 Inserts with broken bosses

Figure 6.31 shows the thermal analysis of two tips being used to heat the insert to be staked in place. The top image shows a tip that just touches the top surface of the insert, and the bottom image one that has a piece going into the hole of the insert. The top one shows very little heat transfer to the insert and its temperature is only about 150 °F (66 °C), which is too low for most plastics. The bottom one shows the insert temperature as about 325 °F (163 °C), which is sufficient to soften most plastics. The extension to the tip also provides positive alignment to the axis of the insert.

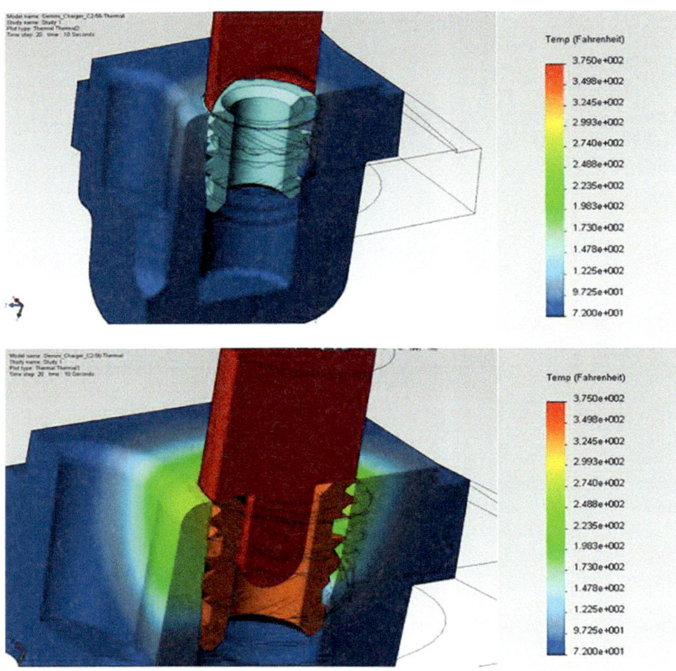

Figure 6.31 Thermal analysis of insert and surrounding area temperature [3]

6.3.2 Insert Height and Required Torque

Two final topics of importance are:

- Height of the insert relative to the surface of the boss that it sits in, and
- The torque applied to the assembly

The insert needs to have its top surface flush with the boss surface. If it is below the surface, the insert will be pulled out or "jack out" when a screw is fastened to it, causing it to loosen itself in the boss. If it is too high, it will interfere with the tolerance of the total assembly height. Since the manufacturing process requires some tolerance, a good practice may be to have it flush to 0.010 in (0.25 mm) below the surface. This is a distance that the insert may be safely pulled without the boss losing its grip on it.

Figure 6.32 shows a case of overtorqueing and jack out. From left to right: the head of the screw left a large depression on the bottom of the counter bore due to overtorqueing. The middle shows the insert being pulled out of its boss. The figure on the right shows the pulled insert leaving its impression on the mating boss. The jack out occurred because the opening in the mating boss was large enough to pull the insert out into itself. It took about 15 kg of force to dislodge the insert from the mating component.

Figure 6.32 Overtorqueing and jack out

6.4 References

[1] Courtesy of Emerson Automation Solutions, *http://www.emerson.com*; Branson Ultrasonics, http://*www.BransonUltrasonics.com*

[2] Courtesy of Dukane Intelligent Assembly Systems

[3] Courtesy of Joseph P. McFadden Sr.; *http://McFaddenCAE.com*

[4] Courtesy of Sonitek

[5] *http://www2.dupont.com/Plastics/en_US/assets/downloads/processing/198118E.pdf*

[6] Source: Jordan Rotheiser; *Joining of Plastics, Handbook for Designers and Engineers*, Hanser Publishers, Munich (2009)

7 Part and Tool Costs

7.1 Introduction

One of my professors used to say that a well-designed part should not only meet or exceed its functional requirements, it should also be optimized for cost.

At Motorola, performance on an individual project was judged on three "ons":

- On spec (specifications)
- On time
- On budget (both development and product costs)

In addition to fulfilling their own responsibilities, most product and program managers are depending on accurate estimates of part and tool costs from the engineers for preparing their business cases for approval of the funding of the programs.

Finally, how does one know that he or she is paying a reasonable price for the product he or she is responsible for? And how does one have data-based negotiations?

With all of the above in mind, let us discuss some basics of estimating part and tool costs.

7.2 Part and Assembly Costs

7.2.1 Elements of a Plastic Part Cost

Let us start with what makes up the cost of a plastic part or assembly. This consists of:

- Cost of the molded part including the procurement, handling, and carrying costs of the resin used
- Cost of bought parts used in the assembly including procurement, handling, and carrying costs

- Cost of secondary operations
- Profit

7.2.1.1 Cost of Molded Parts

This consists of:

- Cost of resin
  - (Part + sprue, runners, and gates weight) × $/kg
  - Procurement, handling, quality control, and carrying costs
- Cost of molding
  - Cycle time × established press and operator rate including overheads (For smaller parts, this is the biggest single contributor to the overall cost. Proper design, tooling, and processing have a major effect on the cycle time.)
- Cost of setting up the mold and establishing steady state (This includes the cost of set up personnel time, parts produced and scrapped until steady state is established, and quality control operations to authorize production quantities.)
- Packaging costs for internal handling and storage, and shipping to the customer
- Cost of molding scrap

7.2.1.2 Cost of Bought Parts

This consists of the cost of parts bought for the assembly, such as inserts, screws, labels, paint, etc. The components of this cost are:

- Actual cost of parts
- Procurement, handling, quality control, and carrying costs

7.2.1.3 Secondary Operations

The cost of each secondary operation consists of:

- Cycle time × established equipment and operator rate including overheads
- Cost of setting up the operation and establishing steady state (This includes the cost of set up personnel time, parts produced and scrapped until steady state is established, and quality control operations to authorize production quantities.)
- Cost of scrap
  - Cost of material + labor expanded × scrap rate

7.2.1.4 Profit

Businesses exist to make money or profit. While they are undeniably entitled to this, it is important to make sure that the profit is:

- Reasonable
- In line with the competition

- Based on the volume of business done–larger customers are entitled to being charged smaller profits because of the volume and dependability of their business
- Totally transparent–no hidden profits

Margin and Mark Up

It is important to distinguish between the terms margin and mark up. Before getting into the details, let us discuss the term "Cost of Goods Sold" or COGS. Simply put, COGS is the actual cost to produce and ship the product including the overheads.

Margin is the amount generally used to add profit to the product and it is defined as the percentage of the selling price that is profit. Thus, if M is the margin, P is the selling price, and C is COGS:

$M = (1 - C/P) \times 100$

To get the selling price (P), divide the COGS by (1 − margin/100):

$P = C / (1 - M/100)$

If the cost (COGS) is \$100, for a 5% margin the selling price would be \$100/0.95 = \$105.26. With a 10% margin, the selling price would be \$100/0.90 = \$111.11, and so on.

Mark up, on the other hand, is simply a percentage of the cost as profit. The same product with a 5% mark up would have a selling price of \$100 + 0.05 × \$100 = \$105. Thus,

5% margin = 5.26% mark up

10% margin = 11.11% mark up, and so on.

7.2.2 Overall Cost to Customer

The overall cost to the customer is equal to the sum of:

cost of (bought parts + molded parts + secondary operations + scrap) + margin

Margin is sometimes built into the shop rates. A familiar example is the \$/h charged by an automotive repair shop. No profit is added at the end of your bill, as it is built into the shop rate.

7.2.3 Determination of Shop Rates, Overhead Costs on Resin and Bought Parts, and Margin

Shop rates and margins can vary by region and business specialty. The molding press rates depend on their size. Thus, a 100 ton press may have a rate of \$35 per hour and a 250 ton press may be \$50 per hour in the U.S. The corresponding rates may be 30 percent of these or less in China.

A typical rate for an ultrasonic machine with an operator may be \$25 per hour. An operator rate may be \$15 per hour.

The procurement team should establish these rates, margins, and overheads on bought items with the selected vendors on a competitive basis. This data should be made available to the personnel doing the cost study.

7.2.4 Determination of Cycle Times

The most critical item is the determination of the cycle time. This is based on the thickest region in the part. Most resin suppliers provide a general approximate chart on thickness vs. cooling time as shown in Figure 7.1. This is a typical chart for a polycarbonate resin. Thus the cycle time can be reasonably estimated.

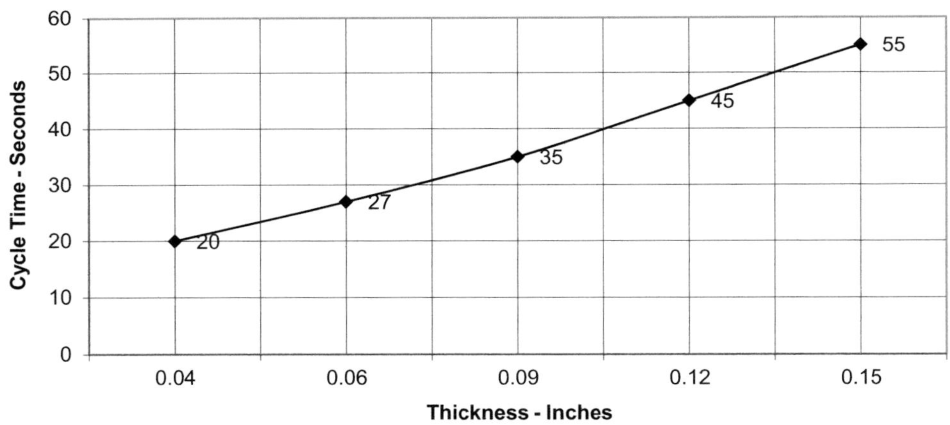

Figure 7.1 Typical estimated thickness vs. total cycle time for a polycarbonate resin

7.2.5 Putting It All Together

The spreadsheet shown in Figure 7.2 puts it all together. The parts being costed are top and bottom covers for a mobile device approximately 100 mm wide, 225 mm long, 13 mm high, and 2.25 mm thick made out of polycarbonate with an approximate density of 1.2 g/cm^3. The top cover has 4 ultrasonic inserts heat staked into it and the bottom has a label applied to it.

The projected area for both covers is 22,500 mm^2 (36 square inches) each. Using the rule of thumb of about 4 tons per square inch, the press tonnage required for each part is 144 tons. Allowing for some additional tonnage for the runner, a 150 ton press is selected for the top cover. Since the bottom cover is a two cavity mold, the tonnage requirement will be doubled and we select a 300 ton press.

Here are the details of the spreadsheet in Figure 7.2:

Cell A2 indicates the quantity per run. This is used to estimate the cost of setup per assembly.

Most of the rest of the cells should be self-explanatory and represent realistic values. Some cells that need explaining are:

Cells F12 and F13: These indicate the number of cavities in each tool. The top cover has 1 cavity and the bottom, 2. This means that processing and runner costs are equally shared by the two cavities, reducing the overall cost of the molded part.

Cells M12 and M13 represent the cycle time taken from Figure 7.1.

Cells J19 and J20 are of special interest. Every time a part is rejected in the respective operations, a corresponding cover is also scrapped. The total secondary scrap needs to include these scrapped parts.

Rows 24-26 and the chart to the right have the total make of the part. Now let us look at cell A25. This is the price the molder quoted you. Cell K25 shows the cost the spreadsheet came up with. Cell L25 shows the difference between the two. This is very powerful. With this information an explanation can be sought from the supplier on the difference. Thus the negotiation now is based on data and not opinions.

An actual working spreadsheet can be downloaded from *www.vikpedia.org*.

Quantity/Run	Overhead on Bought Parts (Percent)	Overhead on Resin (Percent)	Profit Margin
5,000	5	5	10
	$ 0.011	$ 0.042	$ 0.266

1. Cost of Bought Parts

Description	Cost per Piece	Number Needed	Total Per Assembly without
Ultrasonic Insert	0.03	4	0.120
Label	0.10	1	0.100
Total			0.220

2. Cost of Molded Parts

Part	Minimum Machine Size	Part Weight (g)	Sprue and Runner Weight (g)	Total Weight per Part (kg)	No. of Cavities	Setup Charge ($)	Material	Price/kg ($)	Press Rate ($/h)	No. of Operators (if any)	Operator Rate ($/h)	Total cycle (seconds)	Scrap Rate (%)	Run Quantity	Cost w/o Scrap / Part	Scrap Cost per Part	Total Assy Cost inc. Scrap
Top Cover	150	70.31	10.00	0.080	1	150	PC	5.00	40.00	1.00	15.00	40	2.0	5,000	1.04	0.02	1.06
Bottom Cover	300	81.21	11.000	0.087	2	150	PC	5.00	55.00	1.00	15.00	40	3.0	5,000	0.84	0.03	0.86
Total															1.88	0.05	1.93

3. Cost of Secondary Operations

Operation	Cycle	Machine Used	Machine and Operator	Set Up	Run Quantity	Cost of Operation	Scrap %	Cost of Parts Scrapped	Total Cost of Operation
Ultrasonic Inserting	10	U.S. Machine	25	50	5,000	0.079	5.0	1.06	0.13
Label	10	N/A	15	0	5,000	0.042	2.0	0.86	0.06
Total						0.121			0.192

4. Overall Cost Make Up

Price Charged	Plastic Parts	Bought Parts	Secondary Operations	Molding Scrap	Secondary Scrap	Overhead on Bought Parts	Overhead on Resin	Total COGS	Profit Margin	Total Accounted for	Unexplained
3.00	1.88	0.22	0.12	0.05	0.07	0.01	0.04	2.39	0.27	2.66	0.34
100%	63%	7%	4%	2%	2%	0%	1%	80%	9%	89%	11%

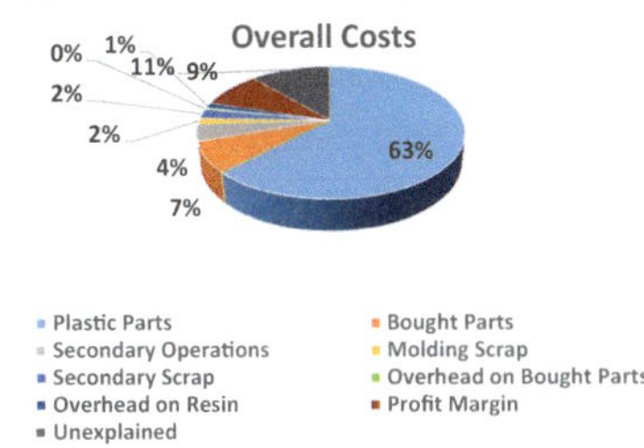

Figure 7.2 Breakdown of total plastic assembly cost

7.3 Tool Costs

While part costing is almost a science, this is not the case with tool costing. In my experience, most of the quotations are based on "seat of the pants" estimates. The estimates also are based on recent similar parts. Another factor is "how hungry" the toolmaker is. In lean times prices will come in very low and vice versa.

It is not at all unusual to see quotes vary by more than fifty percent or more for exactly the same molds. How does one determine what is the right price, then?

The key is to have

a) Well-defined specifications, and

b) A detailed breakdown of the cost

While a) is obvious, b) encourages the toolmaker to pay a little more attention to data-based costing.

7.3.1 Tool Specifications

Having a fully filled specification for the tool being quoted does the following:

a) The customer knows exactly what he or she is getting

b) The customer can perform an apple to apple comparison between quotes from different suppliers

The customer can also use it as a checklist at "delivery."

Figure 7.3 shows a typical mold specification sheet. Note that it also has the part material type and shrinkage. This makes sure there is no error in designing the mold to the right shrinkage.

MOLD TYPE
- [ ] 2 - Plate
- [ ] 3 - Plate
- [ ] Multicavity [ ] Number
- [ ] Family
- [ ] Stripper Plate
- [ ] Hot Sprue Bushing and Manufacturer
- [ ] Hot Runner and Manufacturer
- [ ] Controller Type Included Yes [ ] No [ ]
- [ ] Multishot (Rotational Mold)
- [ ] Reverse Injection
- [ ] Other

CAM ACTUATION
Number of Cam Actions [ ]
- [ ] Mechanical
- [ ] Lifter
- [ ] Collapsible Core
- [ ] Hydraulic Cylinder
- [ ] Other

GATE & RUNNER TYPE
- [ ] Direct Sprue
- [ ] Edge
- [ ] Tunnel
- [ ] Pin Point (Hot Tip)
- [ ] Full Round
- [ ] Half Round
- [ ] Trapezoid

Specify Manufacturer
- [ ] Other

Shrinkage Value (in %)
Direction of Flow [ ] Cross Flow [ ]
Material Type and Grade
Approximate Mold Size
Other Features Not Captured Above

MOLD BASE STEEL TYPE
- [ ] Low - Medium Carbon Steel
- [ ] P20
- [ ] Other

MOLD FEATURES
- [ ] Guided Ejection
- [ ] Straight or Interlocks
- [ ] Vent Pins or Vacuum Lines
- [ ] Locating Ring
- [ ] Other

COOLING LOCATION
- [ ] Cavities
- [ ] Cores
- [ ] Cams
- [ ] Other
- [ ] Mold Plates

TYPE OF COOLING
- [ ] Water
- [ ] Oil
- [ ] Cartridge Heater

WEAR PREVENTION
- [ ] Wear Plates
- [ ] No. of Plates

Material
- [ ] Lubrication Fittings
- [ ] Grease Rings/Grooves

MATERIAL TYPE & HARDNESS
- [ ] Cavities: RC
- [ ] Cores: RC
- [ ] Gate Inserts
- [ ] Shut-Offs
- [ ] Temper (RC) Core Cav
- [ ] Slides
- [ ] Gibbs

EJECTION METHOD
- [ ] Ejector Blades
- [ ] Unscrewing
- [ ] Stripper Plate
- [ ] Air Poppet
- [ ] Ejection from Core Side
- [ ] Ejection from Stationary side
  - [ ] Hydraulic
  - [ ] Electric
  - [ ] Mechanical

FINISH & COATINGS
- [ ] SPI/SPE # - Cavity
- [ ] SPI/SPE # - Core
- [ ] SPI/SPE # - Other
- [ ] Nickel
- [ ] Chromium
- [ ] Texture

OTHER FEATURES
- [ ] Date Clock
- [ ] Revision Insert

Figure 7.3 A comprehensive mold specification sheet

7.3.2 Tool Costing Sheet

Table 7.1 shows a comprehensive mold cost breakdown sheet. Mold manufacturers can be required to submit this sheet along with the tool quotation. If there are knowledgeable tooling engineers within the organization, they can use this sheet to make sure the prices being charged are reasonable. Even in the absence of such personnel, it can provide a very useful tool to compare quotes from two or more suppliers.

Table 7.1 A Comprehensive Tool Costing Sheet

Labor and Processing	Hours	Rate	Total	Comments
Initial Analysis and Mold Design				
CNC Programming				
CNC Machining				
Electrode Manufacturing				
EDM Machining				
Other Machining				
Heat Treating				
Grinding and Fitting				
Polishing and Handwork				
Texturing				
Plating				
Final Mold Assembly				
Transportation				
Mold Trial				
Part Inspection				
Capability Study				
Other				
Total				
Mold Materials				
Mold Base				
Other Mold Components				
Hot Runner System				
Other				
Total				
Molding Trial(s)				
Resin				
Inserts				
Other Components				
Other				
Total				
Grand Total				

8 Six Sigma Techniques in Plastics

8.1 Introduction

The Six Sigma methodology for improving processes and designs has been around since the eighties. As a firm believer in the methodology, I have successfully used it to improve products and performance on multiple projects, saving millions of dollars in the process. I will also be the first to say that in a lot of cases Six Sigma has been used more as a marketing ploy—"We are a Six Sigma company!"—than out of a real desire to improve products or processes.

The following says it all.

Be it as it may, here is a quick introduction to Six Sigma—especially as related to the improvement in design and manufacturing of plastic parts.

8.2 Brief History

Motorola decided to provide extra margin to their customers to produce more robust products. They did this by controlling their processes so that the standard deviation, σ, of their processes was 1/12 of the difference between the USL (upper specification limit) and the LSL (lower specification limit), and the drift of the process mean from that specified was limited to no more than 1.5σ.

When the process met these criteria, the fraction of parts falling outside the range from the USL to the LSL was at most 3.4 ppm, almost a 1,000 times improvement over the traditional approach. This was the heart of the Motorola 6σ approach.

Referring to Figure 8.1, the process variation is reduced to half the design tolerance (i.e., Cp (overall process capability) = 2.0) while allowing the mean to shift as much as 1.5σ from the target. The area under the shifted curves beyond the six sigma ranges (i.e., the tolerance limits) is only 0.0000034, or 3.4 ppm.

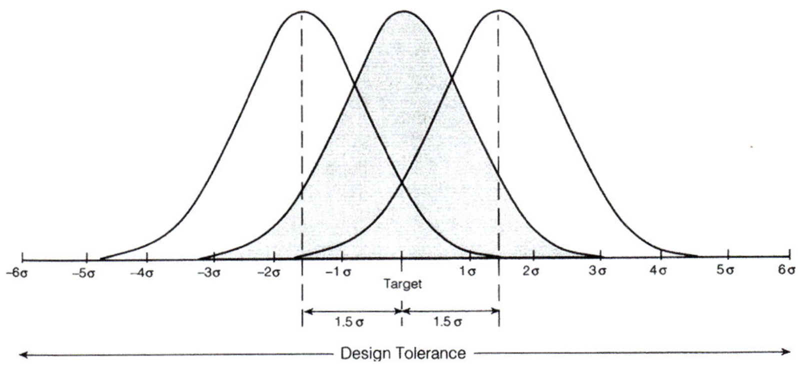

Figure 8.1 Standard deviation and tolerance relationship to achieve Six Sigma

8.3 Evolution of Six Sigma

The Six Sigma methodology has evolved over the years and today it is the structured application of tools and technique applied on specific projects in order to achieve sustained strategic results.

It is aimed at the "big Y's" or strategic goals. As an example,

profit improvement (Y) = $f(x_1)$, $f(x_2)$, $f(x_3)$, ...

where $x_1, x_2, x_3, \ldots$ may represent, respectively, selling cost, product cost, warranty cost, and so on.

The main methodologies are DMAIC and DMADV, and they apply to *all* functions and product development and not just manufacturing as in traditional yield improvement techniques. It is based on a team approach and it includes people in authority and with responsibility along with other members of the team, including experts formally trained in the methodology and subject matter experts (SMEs).

Figure 8.2 illustrates this evolution in terms of the business impact, going from the literal definition to an enterprise-wide philosophy.

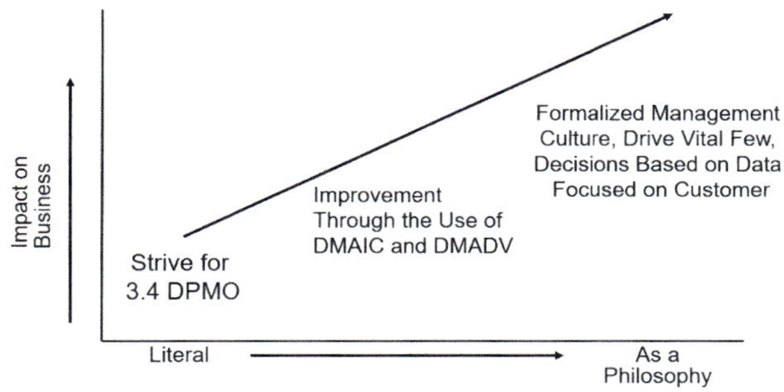

Figure 8.2 Evolution of Six Sigma

In the traditional methods, a group of professionals gets together to resolve problems as they occur. They may or may not correctly address the root cause. Ford 8D[1] is a more disciplined approach that seeks to find the root cause and a permanent data-based solution. This brings the process back to its original state, and generally does not improve the process in any substantial way. When the next problem occurs, the same method is repeated.

The Six Sigma DMAIC (DMAIC and DMADV to be explained shortly) method results in a continuous improvement over time. The DMADV completely redesigns the process, resulting in a radical improvement over a shorter period of time.

Figure 8.3 shows the speed and permanency of traditional versus Six Sigma methods of improving processes and designs.

Using the railroad example, the traditional method would improve the design of the boiler of the steam locomotive to a more reliable one, eliminating the occasional loss of steam and accidents due to boiler failures. The DMAIC method would continuously improve the root causes of slow and inefficient travel by creating

[1] Eight Discipline method developed at Ford Motor Company, used to resolve problems, typically by engineers or other professionals, and focused on product and process improvement by identifying, correcting, and eliminating recurring problems.

newer, more efficient locomotives, leading perhaps to the electric ones going from 30-40 mph to 70-80 mph, accompanied by fewer breakdowns. The DMADV would completely change the design, going to bullet trains and Maglev trains approaching speeds of 300 mph with almost no breakdowns.

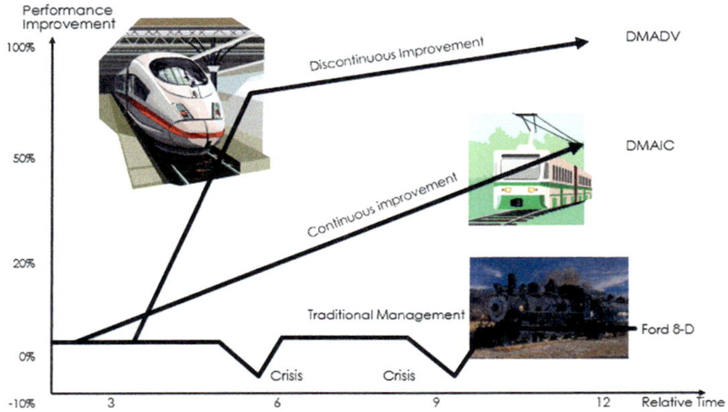

Figure 8.3 Traditional vs. Six Sigma methods [1]

8.4 DMAIC

Engineers responsible for robust designs traditionally do not get involved with process improvements. However, they should be prepared to do this as a part of the holistic approach team.

DMAIC (Define, Measure, Analyze, Improve, Control) is employed to improve an existing process; examples being poor yield, high cost, lack of customer satisfaction, etc. Figure 8.4 breaks down the stages in greater detail.

Let us take the example of poor yield of a plastic part.

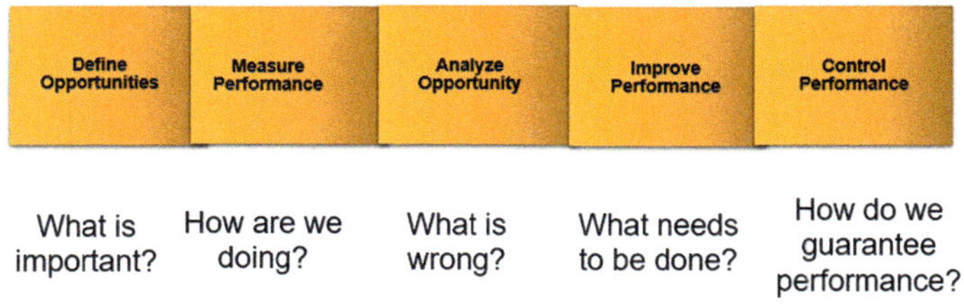

Figure 8.4 The DMAIC method

8.4.1 Define

- Form team of the required personnel as described earlier
- Define team goals, milestones, and budget

As an example, improve the yield from the current 80% to 98% in three months with team expenses of less than $5,000.

8.4.2 Measure

- Measure current conditions

As an example, get yield data from different shifts and different machines for a particular part over a period of time.

8.4.3 Analyze

- Analyze the above data using tools described in detail in the DMADV section and other statistical analysis tools including DOE (design of experiments); systematically find root causes and solutions
- Separate the vital few from the trivial many

As examples,

- The consistently poor yield on a particular machine because of a leaky check valve resulting in inconsistent filling and packing
- Failures due to poor drying of the plastic
  - Find the cause for poor drying, such as poor personnel training, faulty equipment, etc.

8.4.4 Improve

- Improve by implementing solutions based on the above analysis

For example,

- Replace the leaky check valve
- Train the shop personnel in proper drying techniques

It is important to separate the vital few from the trivial many (see discussion in the DMADV section). For example, if there is a rare scratch in a part due to an operator error, a word with the operator in question may be enough, without having to retrain the entire crew.

8.4.5 Control

For example,

- Put in place controls that will proactively check for leaky valve condition
- Put controls in place to make sure drying parameters are continuously monitored
- Put control charts in place to proactively alert of potential out of control conditions

8.5 Design for Six Sigma and DMADV

DMADV (Define, Measure, Analyze, Design, Verify) and DFSS (Design for Six Sigma) are the areas design engineers–being the quarterbacks they are–need to be more proactive in. We will therefore have a more detailed look at this methodology.

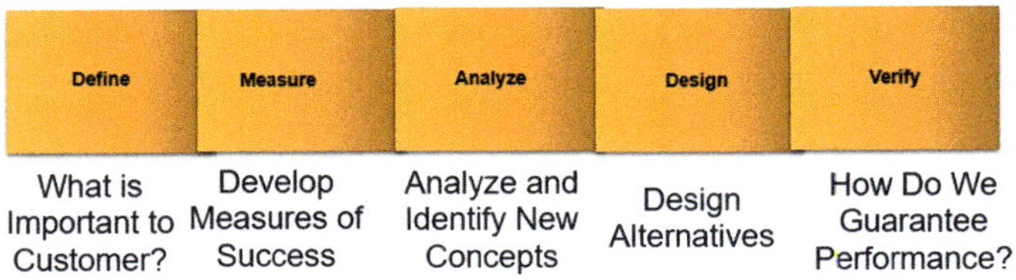

Figure 8.5 The DMADV method

8.5.1 Define

This phase requires that the design team take the time to understand the customers' needs. While the needs of the external customers are paramount, we also need to take into account the needs of the internal customers. As examples,

- The industrial designer needs to keep in constant touch with the mechanical engineer to make sure that the design is practical, that it allows for practical draft and radii, etc. Violate these rules only if absolutely necessary.
- The mechanical engineer needs to make sure that the design is manufacturable and, if the part is an electronic enclosure, that it meets the electrical engineer's needs, and so on down the line.

The process starts with establishing an agreed upon set of CTQs (Critical to Quality). These include the following important requirements:

- Cost
- Minimum performance
- Weight
- Reliability
- Manufacturability

The next step is to translate these needs into measurable/actionable parameters using QFD–Quality Function Deployment (also called the House of Quality)–and obtain consensus on the goals. The QFD process is explained with an example provided below.

The final step is to create a design document that serves as a set of specifications and deliverables for the project, including scope, schedule, and budget.

QFD (Quality Function Deployment)

See Figure 8.6. This example uses an *internal* plastic part in a handheld electronic device. Looking at the columns from the left to the right,

- The first column represents the customer (internal *and* external) needs
- The second, their importance on a scale of 10-0, 10 being very important and 0 being not important at all
- The next four columns represent the relationship of the needs to the factors material, design, processing, and tooling, with the following weightage:
  - 9: Strong relationship
  - 3: Moderate relationship
  - 1: Little relationship
- The bottom four cells on the right represent the multiple of the importance times the respective weightage

Now, looking at the analysis, durability is of course very important. While the material processing and tooling will have some effect on this factor, the most important factor is design, such as avoiding sharp corners, wall thickness variation, etc.

Material has some effect on the weight because of the density and its ability to be molded with a thinner wall. Design again is the most important factor–making the part no thicker than the design needs. Processing and tooling have little effect on this factor.

Cost has some dependence on the material ($/kg) and tooling (reducing cycle time with proper cooling); the two most important factors are design and processing (lowest cycle time, highest yield).

The aesthetics are not as important, as it is an internal part. Glass fibers on the surface and witness lines from the tool are acceptable. However, design and processing still play a key role in it. For example, sink marks, short shots, visible knit lines, flash and its removal, gate vestige are important considerations with both the design and processing having a major contribution. The material and tooling can have some effect on it, such as poor color mixing due to the material and flash due to poorly designed parting lines.

The final count has design with a score of 171 followed by processing (115), material (75), and tooling (61), each as respective contributors to meeting the customer's needs.

		FACTORS			
Customer Needs	**Importance**	**Materials**	**Design**	**Processing**	**Tooling**
Durability	9	3	9	3	3
Light Weight	7	3	9	1	1
Low Cost	6	3	9	9	3
Aesthetics	3	3	9	9	3
Feature Priority		75	171	115	61

Figure 8.6 Quality Function Deployment (QFD)

8.5.2 Measure

The measure phase includes a thorough review (measurement) of the available data. This may include:

- Detailed review of the existing design (if the new product is replacing it)
- Competitive products benchmarking, if this is a new design
- Most useful features of the current design
- Features of the existing design that can be discarded
- Missing features in the current design
- Lessons learned from previous similar designs

Based on this "measurement":

- Establish new specifications–functional, ruggedness, cost, etc.
- Establish and commit to budget and schedule
- Identify potential process issues and their prevention and/or backup options

8.5.3 Analyze

This phase identifies, consolidates, and solidifies concepts for the new design. The process includes challenging assumptions and paradigms. Nothing is off the table unless data shows that it is impractical. Starting with "blue skying," the process narrows down the design choices to a few of the most viable ones.

8.5.4 Design

The next stage is creating the actual design with the most robust solutions. Some of the tools used in this process are:

- Six Sigma based:
  - Failure Mode and Effects Analysis (FMEA)
  - Error proofing, such as Poka Yoke
  - Benchmarking
- Conventional:
  - Tolerance analysis
  - Simulation (flow–Moldflow, Moldex–structural, etc.)
  - Prototype testing

8.5.4.1 Failure Mode and Effects Analysis (FMEA)

This is a very important tool in proactively looking at and designing around possible failure issues. The example below looks at one such analysis for the outside enclosure of a smartphone, such as the one shown in Figure 8.7.

Figure 8.7 Smartphone enclosure

Table 8.1 shows the actual analysis. The headings on the top row are fairly self-explanatory. Let us look at some that may not be.

Severity indicates the effect of the failure on a scale of 10 to 0, 10 being catastrophic and 0 causing hardly any issues. Chance of occurrence, as the name indicates, is the likelihood of the defect occurring based on existing information. The 10-0 scale indicates, from highest to lowest, the probability of the defect occurring. Ease of detection indicates if the failure will be obvious or may not be detectable: fatigue failures may not be detected until the failure actually occurs; similarly, cracks may start in a microscopic form that cannot be detected until they cause a catastrophic failure. The 10-0 scale goes from difficult to easy to detect. RPN stands for Risk Priority Number and it is the multiple of the above three (severity, chance of occurrence, and ease of detection).

A group of users and SMEs together conducted this analysis. Let us look at the highest and lowest RPN issues. The others follow the same logic.

A crack in the enclosure will expose the internal parts; it has a severity of 9 because in most cases the entire enclosure may need to be replaced even if the internal members survive. If the part is designed properly, the most likely cause of the failure is poor drying of the material, as discussed in the previous chapters. The chances of an occurrence are moderate, based on most people being careful in not dropping the phone or protecting it by putting a protective enclosure around it. The score is thus 5. The crack may not be easy to detect, as it may start as a hairline rack on the inside. The total RPN is thus $9 \times 5 \times 9 = 405$—very high.

Table 8.1 FMEA for Smartphone in Figure 8.7

Function of Part	Potential Failure Mode	Potential Effect(s) of Failure	Severity	Potential Cause(s)/ Mechanism(s) of Failure	Chance of Occurrence	Current Process Controls	Ease of Detection (Lower Number Easier to Detect)	RPN
Hold Internal Parts Together	Crack	Internal Parts Exposed	9	Improper Processing (Poor Drying)	5	Controlled Processing	9	405
	Scratch	Aesthetic Appeal	1	Normal Abuse	3	Controlled Scratch Resistance Coating	5	15
Prevent Water Ingress	Seal Failure	Possible Loss of Functionality	5	Exposure to high heat and humidity	3	Seal testing at elevated temperature and humidity	8	120
Provide Aesthetic Appeal	Flow Marks	Poor Consumer Appeal	7	Poor Design	7	Check Design for Errors	3	147

A surface scratch, on the other hand, may only be an irritant that one gets used to in a short time. The prevention is also pretty simple, done by applying a scratch-resistant coating to it. Using the logic based on the previous defect, the RPN is $1 \times 3 \times 5 = 15$. Based on this data, it may be easier for the marketing department to make a call on whether the extra cost of applying a scratch-resistant coating is justified based on competitive factors.

8.5.4.2 Poka Yoke

Poka-yoke is a Japanese phrase that means "mistake-proofing" or unintended error prevention. Toyota Motors has made the most widespread use of this technique in building the world-leading quality of their cars.

These tools can be used both in the products themselves or in the process leading to the design of the products. Process-related Poka Yokes will be discussed a little later.

Figure 8.8 shows a couple of examples of Poka Yoke in real life. The cut corner in the label prevents the placement of a label in the upside down position. This is especially useful today, as the labels may be in a language that the operator does not understand. The second example shows the ubiquitous two-prong electrical plug in the U. S. The wide blade is neutral and the short one "hot." The corresponding slots in the receptacles have corresponding widths; thus, the plug can only go in one way. This prevents an accidental electrical shock to the user of the appliance connected to the plug.

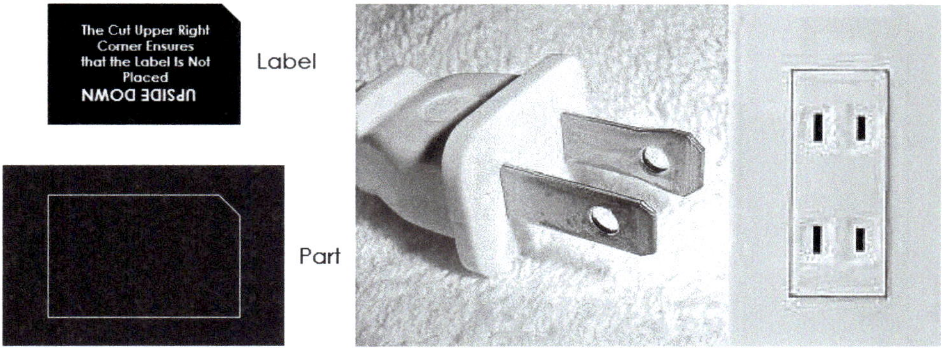

Figure 8.8 Some common Poka Yoke examples

8.5.4.3 Conventional Tools Used in Design Phase: Simulation

Simply put, simulation prevents a lot of costly trial and errors in coming up with the most robust design and process. In the plastics world there are mainly two types:

1. Flow

   The flow of the plastic is simulated, including pressures, temperatures, and thermodynamics of the process. Some of the common modules are Moldflow, Moldex, and Sigmasoft.

2. Structural, also referred to as FEA (Finite Element Analysis)

   The mechanical stresses are simulated both in static and dynamic modes, such as drop to the floor. Some common modules are Abaqus, ANSYS, and Cosmos.

The following basic requirements need to be met in order for the analysis to be accurate and to prevent the use of these rather expensive resources in a wasteful manner:

- The assumed material properties used in the analysis need to be reliable
- The cross-sectional areas and wall thicknesses used in the analysis should not have more than marginal changes remaining before the design is finalized
- If a plastic part is not designed per the established rules, there may be huge molded-in stresses or stress concentrators that the simulation will not capture: (see Figure 8.9) as described in Chapter 3, the huge temperature difference between the adjacent thin (118 °C) and thick (238 °C) sections and their subsequent non-uniform cooling may introduce a high stress at the junction that a mechanical simulation software may not detect
- Substantial changes may be required for proper flow in the wall thicknesses in an improperly designed part that may cause functional and spatial issues with the original design, and may require substantial re-design and repeated simulations

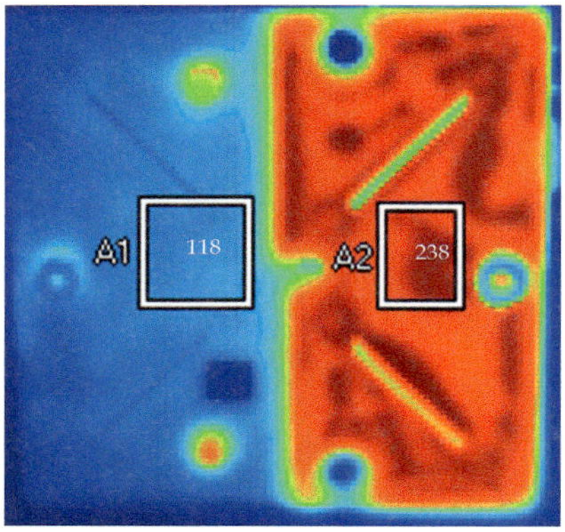

Figure 8.9 High stress at the junction of thick and thin wall [2]

8.5.4.4 Some Common Poka Yoke Tools for the Design Process

8.5.4.4.1 Checklists

Checklists have existed forever. The most common example is the grocery shopping list—a list that prevents not buying a needed item and (hopefully) not buying an unneeded item.

Engineering organizations have created short and long checklists for checking designs based on best practices and past experiences. The issue with these lists is that they are long and in most cases "after the fact." This leads to an incomplete and inefficient use of the lists most of the time.

DFMPro

There is at least one CAD-based software called DFMPro from Geometric, Ltd. that actually checks for design errors in the solid model itself based on established industry and custom design rules as the part is being designed. Figure 8.10 shows the detection by this software, within the model, of potential sink issues that were ignored, resulting in sink marks and shadows. More information about this software will be provided in the reference chapter (Chapter 9).

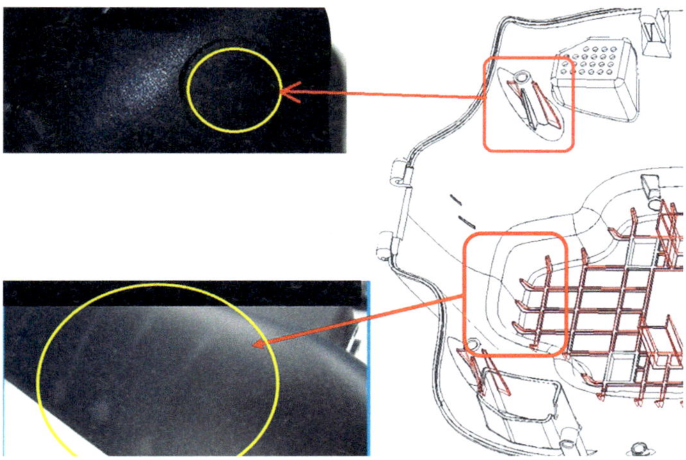

Figure 8.10 Detection of design errors within CAD: sink mark, top left, and shadows, bottom left

8.5.4.4.2 Design Reviews

Design reviews with peers and subject matter experts in an organized and timely manner is another existing design process Poka Yoke tool.

8.5.4.5 Some Other Commonly Used Tools

8.5.4.5.1 Pareto Analysis

The Pareto principle (also known as the 80-20 rule, the law of the vital few and trivial many) states that, for many events, 80 % of the effects come from 20 % of the causes. Business management thinker Joseph M. Juran suggested the principle and named it after Italian economist Vilfredo Pareto, who observed that 80 % of the income in Italy went to 20 % of the population. It is a common rule of thumb in business; for example, "80 % of your sales come from 20 % of your customers."

Figure 8.11 is an analysis conducted on common defects in plastic parts, based on observing one hundred defects. It shows that 70 % of the defects were caused by visual defects, sharp internal corners, and warpage. If we conducted a Pareto analysis of these defects, we would probably find that 2-3 issues, like non-uniform walls and sharp corners, contributed to most of these types of defects. Fixing these on a priority basis would result in a swift and dramatic increase in the yields.

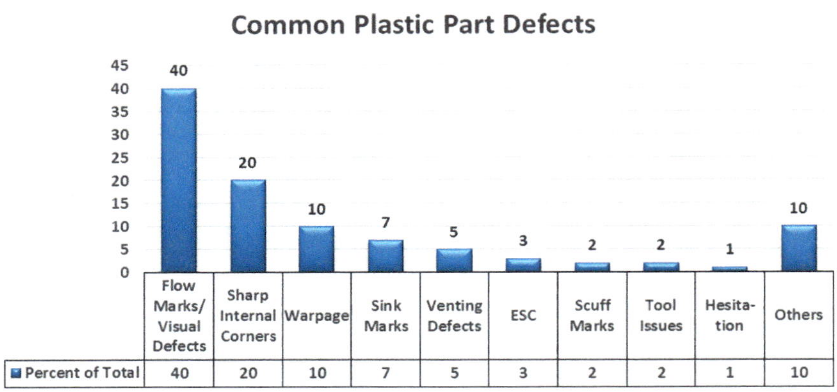

Figure 8.11 Pareto analysis applied to plastic part defects

8.5.4.5.2 Cause and Effect or Fishbone Diagram

Traditionally, this consisted in breaking down the cause of any defect into four "M's": Men, Method, Machine, and Material. In most recent times, two other important categories have been added: Measurement and Environment. Looking at any defect with a group of subject matter experts, and doing it systematically and efficiently, leads to and builds consensus on the main causes of the defects.

Figure 8.12 shows a part with unacceptable flow lines. Figure 8.13 shows a thorough review of the possible causes that could cause the flow lines to appear. After going through all the possible ones, the consensus was that the main contributors were uneven walls, flow from thin to thick, and the pigment quality leading to immiscibility with the resin. The causes were addressed and the defects eliminated.

Sometimes, the agreed upon apparent cause may not be the one contributing to the defect. In this case a study has to be conducted again. However, the new study successfully eliminates some of the more apparently obvious cases.

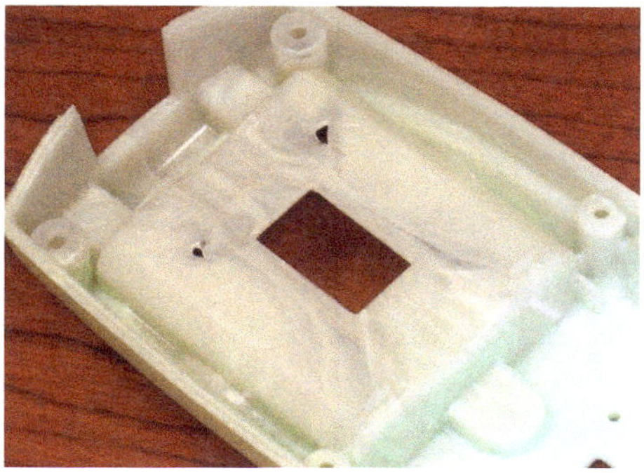

Figure 8.12 Molded part with unacceptable flow lines

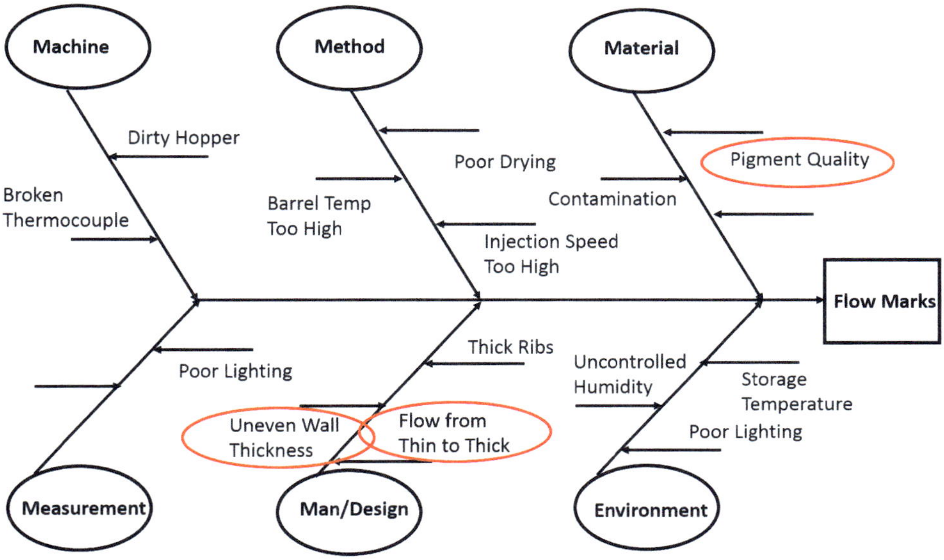

Figure 8.13 Fishbone diagram for causes of flow marks

8.5.4.5.3 Five Whys

This is another very simple and elegant tool that works well with the cause and effect analysis. Asking "why" five times with a probing attitude can often lead to the answer quickly and systematically.

Take for example a part that has a corner broken on impact. Asking "why?" just four times led to the root cause of the problem. The designer did not know that a sharp internal corner leads to a high stress concentration as shown in Figure 8.14 to the right.

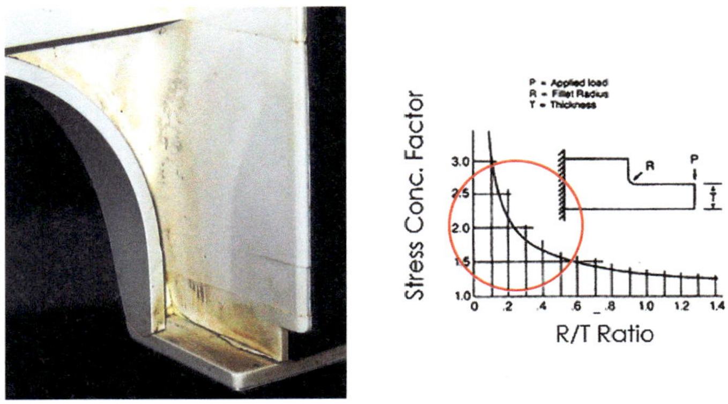

Figure 8.14 Broken corner in plastic part and stress concentration factor

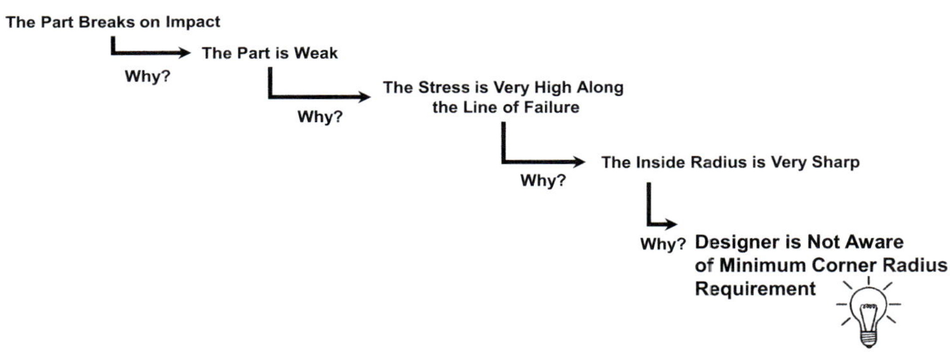

Figure 8.15 Five whys on broken corner

8.5.4.5.4 Design of Experiments

While tools like the fishbone diagram may provide answers to simple situations, they may not be able to do so for more complex ones. Even when they do, they are more qualitative in nature and may not be able to optimize the performance. There may also be a confounding issue where two or more variables interact with each other, and changing just one may not solve the issue.

Let us take the example of baking a cake. In order to get the appearance and taste just right, the temperature and time have a confounding issue. We can try changing the time and temperature one at a time all day and may not get the baking just right. We cannot reduce the temperature and bake it for a longer time or vice versa to get it right. The right combination happens to be 175 °C for 30 minutes. In this case, a simple DOE with two or more variables can lead us to the ideal conditions quickly and reliably. Chapter 5, Section 5.5.3.8, describes an actual DOE study to improve the appearance of a knit line.

8.5.5 Verify

The verification phase makes sure that all the design specifications are met. This should include:

- Verification of the design based on
  - small runs (lab, functional, and beta testing)
  - pilot runs

Additionally, it also confirms that the long-term production process has been established, handed over to the process owner–generally manufacturing–and is running smoothly.

It is important that the design team stay involved until the manufacturing process has matured.

8.6 Going Beyond Just Making Good Products

To conclude this chapter, here are some parting thoughts. In this day and age, "good enough" is not good enough anymore. We need to go beyond the customers' expectations to stay competitive. Let us discuss some tools and topics that have inspired me during my working years.

8.6.1 Kano Analysis

The Kano model looks at product development that results in customer satisfaction; it was developed in the 1980s by Professor Noriaki Kano. This roughly classifies customer satisfaction into must meet needs, needs that either satisfy or dissatisfy, attractive qualities that are liked if present but do not necessarily cause dissatisfaction if not, attributes that customers are indifferent to, and attributes that may please some and displease some–such as the no headphone jack in the Apple iPhone 7.

Figure 8.16 is a Kano analysis for an automobile. The gray box in the middle represents attributes that do not necessarily make the car stand out from its completion. The must needs are all four wheels fitted with proper tires. No one is going to buy a car that has only three wheels. The expected miles per gallon for this car are 25 miles per U.S. gallon (approximately 10 kilometers per liter). Below this will result in lower satisfaction and vice versa. Whether the interior is made of ABS or polycarbonate does not really matter to the customer as long as it meets the basic durability and finish requirements. It may mean higher or lower profit to the manufacturer. Having an automatic anti-collision mechanism in a moderately priced car in 2017 is a "delighter" and will certainly attract customers. Notice that the previous sentence was qualified with "in 2017." As time goes on, delighters become basic requirements. A bluetooth link was a delighter five years ago. It is a basic need today. Having a digital speedometer may please some and displease some.

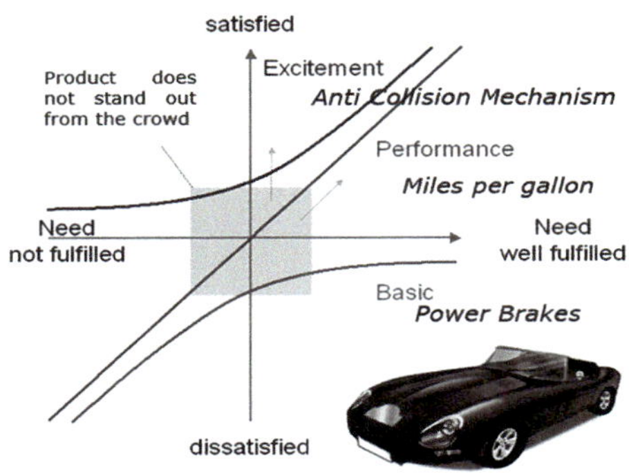

Figure 8.16 Kano analysis for an automobile

Let us think how we can apply this to plastic parts. You are designing an enclosure for a handheld mobile device:

- A highly visible knit line on the front surface would not meet the basic needs
- If the expected weight of similar products is 200 g, a higher weight will take away satisfaction and a lower weight will add more satisfaction
- If the product can be dropped multiple times on asphalt from six feet numerous times without a scratch or a crack, it will be a delighter
- If the product is made out of ABS ($2/kg) or polycarbonate ($5/kg), it will make no difference to an external customer as long as the basic needs are met (your boss may give you a raise though, if it can be made out of the lower cost ABS without sacrificing any needed attributes!)
- If it has a shiny surface instead of the conventional textured surface, it may please some users and displease others like me

8.6.2 Breaking the Paradigm

Figure 8.17 Copernicus 1580 portrait (artist unknown) in the old town city hall, Toruń, Poland

In 1530, Copernicus completed and gave to the world his great work *De Revolutionibus*, which asserted that the earth traveled around the sun. The belief then was that the sun traveled around the earth.

Bruno and Galileo publicly agreed with him. For such blasphemy, Bruno was tried before the Inquisition, condemned, and burned at the stake in 1600. Galileo was brought forward in 1633, and, there, in front of his "betters," he was forced to his knees to renounce all belief in Copernican theories, and was thereafter sentenced to imprisonment for the remainder of his days.

Such was the price for breaking a paradigm.

Figure 8.18 shows the evolution of the light bulb. If it were not for engineers willing to break the paradigm, we would still be stuck with the very low efficiency incandescent bulbs with their less than pleasing light colors. I would not be surprised if many engineers who first came up with the idea were met with a lot of skepticism and ridicule. Some may even have lost their jobs because they did not succeed the first time creating the revolutionary product.

Turning to plastics:

Circa 1965: Making bumpers out of plastic? "A crazy idea that will not work." In 1968, General Motors introduced an "Endura" body-colored plastic front bumper designed to absorb low-speed impact without permanent deformation in the Pontiac GTO. The metal bumpers were gone forever?

An airplane with composite wings? Meet the Boeing 787 Dreamliner!

Figure 8.18 Paradigm shift from incandescent to fluorescent to LED

8.6.3 BHAGS[2] (Pronounced "Be Hags"): Big Hairy Audacious GoalS

Innovation and thought leaders have set goals that seem impossible to achieve[3].

In a special address to the Congress on the importance of space on May 25, 1961, President John Kennedy declared:

"... if we are to win the battle that is now going on around the world between freedom and tyranny, the dramatic achievements in space which occurred in recent weeks should have made clear to us all, as did the Sputnik in 1957, the impact of

[2] An idea conceptualized in the book "Built to Last: Successful Habits of Visionary Companies" by James Collins and Jerry Porras

[3] Figure above: non-copyrighted image from flickr.com

this adventure on the minds of men everywhere, who are attempting to make a determination of which road they should take ...

I therefore ask the Congress, above and beyond the increases I have earlier requested for space activities, to provide the funds which are needed to meet the following national goals:

First, I believe that this nation should commit itself to achieving the goal, *before this decade is out*, of landing a man on the moon and returning him safely to the earth ..."

Figure 8.19 President John F. Kennedy's special address to Congress, May 25, 1961 [3]

On July 20, 1969 (less than ten years after the address), Neil Armstrong and Buzz Aldrin became the first humans to land on the moon [4].

8.7 References

[1] Based on article by Ray Stata; *Organizational Learning: The Key to Management Innovation*, Sloan Management Review, Volume 30, Number 3 (spring 1989)

[2] Thermal image courtesy of John Bozzelli

[3] *https://history.nasa.gov/moondec.html*

[4] *https://www.nasa.gov/sites/default/files/images/589552main_as11-40-5875_full.jpg*

9 Further Reading and Reference Material

9.1 Introduction

"Knowledge is of two kinds. We know a subject ourselves, or we know where we can find information upon it." –Samuel Johnson.

This famous quotation from Johnson dates back to the early 18th century and has become even more meaningful in this age of information overload. I just "Googled" "polymer." It opened 25 pages of results!

The preceding chapters are my personal collection of lessons learned during my more than 46 years as a practicing engineer and they represent "knowing the subject ourselves." This chapter provides information on "where we can find information upon it." For most parts it is divided into the "four wheels" and additional areas supplementing the wheels.

9.2 Materials

By far the best resources for information on plastic resins are their suppliers. Most of them have excellent websites on material properties, design guides, processing guides, and case studies. They can be excellent resources for review and optimization of customer parts. However, this service is gradually becoming available only for really large customers. If you are fortunate to be a large customer, by all means use it. If not, it may not be a bad idea to consolidate and use as many materials from one supplier to get as much leverage as possible.

There is almost no supplier that has all the resins to meet the vast variety of functional needs of components within products. For this reason, it is a good idea to standardize on the less than a dozen or so mostly used types of material that meet most of the needs from the approximately **100,000** resins that are currently available and whose number is growing.

Depending on the nature of the products, some examples are:

- Materials for enclosures or body parts, such as PC, PC/ABS, ABS, PPO, high impact polystyrene (HIPS), glass-filled nylons, PVC, etc.
- Materials for gears, moving parts, and high fatigue requirements, such as nylons, acetals, etc.
- Materials for optical requirements, such as polycarbonates, acrylics, etc.
- Materials for electrical components, such as polybutylene terephthalate (PBT), polysulfones, etc.
- Materials for high temperature resistance, such as polyetherimide (PEI), liquid crystal polymers (LCP), polysulfones, etc.
- Thermoplastic elastomers for standalone or overmolding needs, such as TPU, TPV, SEBS, etc.
- Commodity materials, such as polystyrene, polyethylene, polypropylenes, etc.

An additional important advantage in standardizing the materials is being able to develop an intimate understanding of their strengths and weaknesses both in performance and processing. This allows proactive preparation for their shortcomings in design, tooling, and processing.

Any material search is a two-step process. The first step is to narrow down the candidates through an exhaustive search based on the desired properties. The second step is to finalize the materials search by going to the selected materials websites and/or contacting the supplier to get more advanced information, including design and processing assistance.

9.2.1 Step 1: Search across Suppliers

For the first step of the search, one of the most useful resources is Prospector®. It has currently a database of over 100,000 materials, which is continuously being updated. Its search features include more than 500 searchable properties to easily locate specific properties needed. These include ASTM, ISO, test standards, flammability properties, agency ratings, general requirements, and more.

As an example: the need is to find a Nylon 6 resin with a better than HB flame rating, available in North America, and with a certain desired tensile strength. Prospector has filters that can search for candidates with availability in North America, Nylon 6, HB rating, and tensile strength (in this case between 65 and 90 MPa).

See Figure 9.1 for the results. Starting from 119,116 available resins, the list is quickly boiled down to 244 materials. At this point it can be further refined with other criteria and/or contacting the preferred supplier(s) to help narrow the search to one or more materials that will meet all the required needs, including cost and availability. Also, to help with this process, there is a full datasheet for each search result that includes the supplier and distributor contact information.

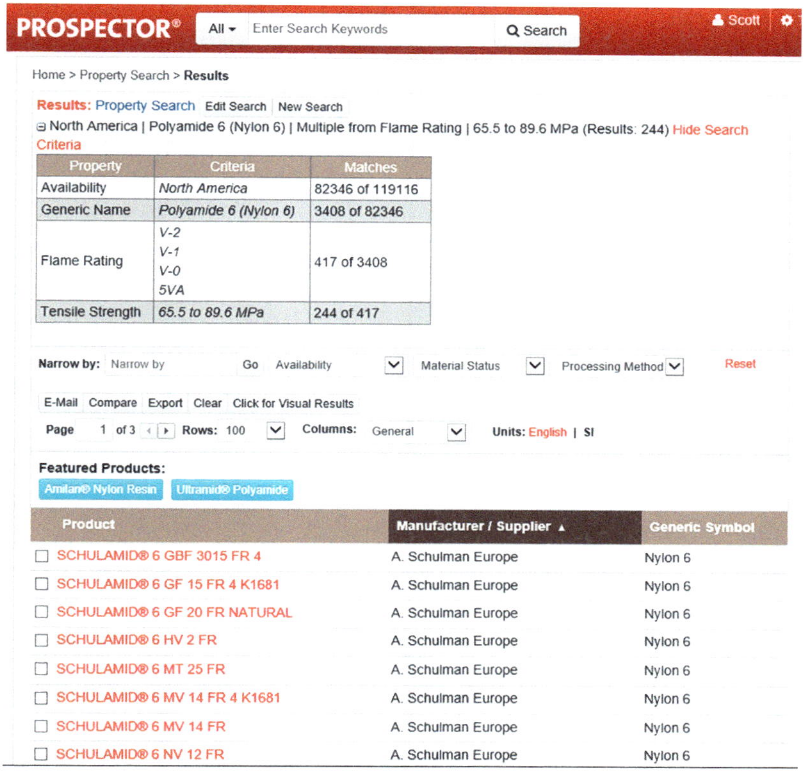

Figure 9.1 Material search results based on customer requirements [1]

Other features include multipoint data and side by side comparisons of various materials for the desired properties. Prospector has both a free and a paid version based on the extent of search that is needed. Below is the link:

https://www2.ulprospector.com/prospector/

9.2.2 Step 2: Search within Suppliers

Most resin suppliers have websites that are goldmines of information. Some of the ones that I have used extensively are:

Covestro (formerly Bayer): http://www.products.covestro.com/en.aspx

SABIC (formerly GE Plastics): https://www.sabic-ip.com/gep/Plastics/en/ProductsAndServices/ProductsAndServices/products_and_services.html

DuPont: http://www.dupont.com/products-and-services/plastics-polymers-resins.html

GLS Thermoplastic Elastomers (now part of PolyOne): http://www.polyone.com/products/thermoplastic-elastomers

9.2.3 RoHS, REACH, UL, and Other Standards

9.2.3.1 RoHS

RoHS [2] stands for Restriction of Hazardous Substances. RoHS is also known as Directive 2002/95/EC, originating in the European Union and restricting the use of six hazardous materials found in electrical and electronic products. All applicable products in the EU market after July 1, 2006 must pass RoHS compliance. It specifies maximum permissible levels in any product for the following ten restricted materials:

1. Lead (Pb): < 1000 ppm
2. Mercury (Hg): < 100 ppm
3. Cadmium (Cd): < 100 ppm
4. Hexavalent chromium: (Cr VI) < 1000 ppm
5. Polybrominated biphenyls (PBB): < 1000 ppm
6. Polybrominated diphenyl ethers (PBDE): < 1000 ppm
7. Bis (2-ethylhexyl) phthalate (DEHP): < 1000 ppm
8. Benzyl butyl phthalate (BBP): < 1000 ppm
9. Dibutyl phthalate (DBP): < 1000 ppm
10. Diisobutyl phthalate (DIBP): < 1000 ppm

9.2.3.2 REACH

REACH (EC 1907/2006) [3] aims to improve the protection of human health and the environment through the better and earlier identification of the intrinsic properties of chemical substances. This is done by the four processes of REACH; namely, the registration, evaluation, authorization, and restriction of chemicals. REACH also aims to enhance innovation and competitiveness of the EU chemicals industry.

The REACH Regulation places responsibility on industry to manage the risks from chemicals and to provide safety information on the substances. Manufacturers and importers are required to gather information on the properties of their chemical substances, which will allow their safe handling, and to register the information in a central database in the European Chemicals Agency (ECHA) in Helsinki. The Agency is the central point in the REACH system: it manages the databases necessary to operate the system, co-ordinates the in-depth evaluation of suspicious chemicals, and it is building up a public database in which consumers and professionals can find hazard information.

The Regulation also calls for the progressive substitution of the most dangerous chemicals (referred to as "substances of very high concern") when suitable alternatives have been identified.

One of the main reasons for developing and adopting the REACH Regulation was that a large number of substances have been manufactured and placed on the market in Europe for many years, sometimes in very high amounts, and yet there is insufficient information on the hazards that they pose to human health and the environment. There is a need to fill these information gaps to ensure that industry is able to assess hazards and risks of the substances, and to identify and implement the risk management measures to protect humans and the environment.

Having entered into force in 2007, REACH provisions are being phased-in over 11 years. Companies can find explanations of REACH on the DG Enterprise or ECHA websites, and can contact national helpdesks.

9.2.3.3 UL 94 Standard for Tests for Flammability of Plastic Materials for Parts in Devices and Appliances

If products need a UL Safety Certification such as UL 60950-1 – Information Technology Equipment - Safety, the part walls require a minimum thickness in order to meet the flammability requirement known as UL 94 [4]. These are intended to serve as a preliminary indication of the acceptability of the material. The ratings range from 5VA to HB (least to most flammable) based on the rate of burning.

More information can be found at: http://ulstandards.ul.com/standard/?id=94_6

9.3 Design

Material suppliers again have excellent literature, even entire books on design with their materials. Here are some excellent resources that I have used:

9.3.1 Structural Calculations Including Snaps and Gears

For primarily amorphous materials, the "book" that I have most often used is from Bayer Material Science (now Covestro), called "Part and Mold Design." This is a goldmine of information on a variety of topics including design of mold elements. Chapter 3 provides almost all the structural calculation basics and formulas that a plastic part designer needs [5]. It can be downloaded from:

https://www.plastics.covestro.com/Library/Technology-brochures.aspx

For semicrystalline materials such as nylon and acetal and gear design, I recommend "Start with DuPont" from DuPont:

http://www.dupont.com/content/dam/dupont/products-and-services/plastics-polymers-and-resins/thermoplastics/documents/General%20Design%20Principles/General%20Design%20Principles%20for%20Engineering%20Polymers.pdf

For snap fit designs:

From BASF:

http://www8.basf.us//PLASTICSWEB/displayanyfile?id=0901a5e1801499d2

For quick structural calculations:

http://www2.basf.us/businesses/plasticportal/pp_techRes_tools_snapfit_en.html

9.3.2 High Performance Optical Parts

Lenses for high-end optics require molded parts with minimal molded-in stresses, in addition to the normal optical properties. These stresses cause unwanted change in direction for the rays entering and exiting the lens. There are specialty materials and molding techniques to achieve this. The following articles, "Practical Plastic Optics" and "Top Tips for Optical Moulding," from Mike Schaub and Sumitomo, respectively, describe the requirements in detail:

https://wp.optics.arizona.edu/optomech/wp-content/uploads/sites/53/2016/10/Plastic-Optics.pdf

https://www.sumitomo-shi-demag.eu/uploads/tx_braindownloads/InjectionWorld_Optical.pdf

9.3.3 Living Hinge

Living hinges are ubiquitous. Successful living hinge design requires an optimal combination of material, design, tooling, and processing to achieve the desired performance. "Design Issues on Living Hinges," below, is a good guide:

http://web.mit.edu/2.75/resources/random/Living%20Hinge%20Design.pdf

9.3.4 Brittle and Ductile Failures

Although all failures need to be avoided within reason, brittle failures can be a lot more catastrophic than a ductile one. An ordinary drinking cup made of glass will have a catastrophic brittle failure when dropped on a ceramic floor. The same cup, made out of polycarbonate, dropped in a similar fashion will at most develop stress whitening marks, indicating the possible start of a ductile failure.

Brittle failures show no apparent plastic deformation before fracture. Ductile failures, on the other hand, show considerable plastic deformation before the failure occurs.

These failures depend on the material, design, processing, rate of loading, and temperature for a given material. Polycarbonate parts are tough and will generally fail in the ductile mode. However, high strain rates (as in falling to the ground from a height) combined with extremely cold temperature, sharp inside corners, and poor drying may cause even a part made out of PC to fail in the brittle mode.

The following link provides a good explanation of these failures:

https://en.wikipedia.org/wiki/Fracture

9.3.5 Thin Wall Molding

Chapter 3, Section 3.4, discussed thin wall, structural foam, and gas assisted molding. Besides design, thin wall presents different tooling and processing challenges, as the plastic pressure requirements are much higher than the required for conventional wall molding. Below is a link to an article on the subject, which discusses these challenges:

http://www.ptonline.com/articles/secrets-of-successful-thin-wall-molding

9.3.6 Structural Foam

Just as in thin wall molding, design, tooling, and processing requirements are different. Below is a comprehensive guide on this process from GE Plastics (now SABIC):

https://www.fmcorp.com/downloads/designguide.pdf

9.3.7 Gas Assisted Molding

Gas assisted molding also has different tooling and processing requirements. Below again is a comprehensive guide on this process from GE Plastics:

http://www.mack.com/wp-content/uploads/2015/03/Gas-Assist_Injection_Molding.pdf

9.3.8 Two Shot Molding

As discussed in previous chapters, two shot molding allows the molding of one resin over another within the same molding cycle or inset molding a substrate with an overmold. Below is the link to a comprehensive guide on the design, tooling, and processing from PolyOne:

http://www.polyone.com/files/resources/Overmold_Design_Guide.pdf

9.3.9 Pad Stamping

Pad stamping allows the transfer of multicolored graphics even onto such complicated surfaces as golf balls. See Figure 9.2.

Figure 9.2 Multicolor pad printed golf ball [6]

Below are links to some comprehensive articles on the process.

http://www.decotechgroup.com/library/pad-printing/tech-bulletin-pad-print-101/

http://www.inkcups.com/blog/pad-printing/41-things-cool-things-about-pad-printing/

9.3.10 Plating

Plastics can be plated with metals both for functional reasons as having a conductive surface and also to provide them with a metallic look. Both the design and resin have to be refined in order to have a plated surface that is both visually appealing and has proper adhesion for wear. The following is a comprehensive guide on the subject from Dow Chemicals:

http://msdssearch.dow.com/PublishedLiteratureDOWCOM/dh_0937/0901b8038 0937128.pdf?filepath=elecmatls/pdfs/noreg/886-00065.pdf&fromPage=GetDoc

9.3.11 In-Mold Decoration

In-mold decoration allows the automatic, robot, or operator assisted application of wear resistant graphics onto plastic parts during the injection molding cycle. Below is a comprehensive guide to in-mold decoration from SABIC:

https://www.sabic-ip.com/gep/Plastics/en/ProductsAndServices/IMDHome/in_mold_decoration.html

9.3.12 Ultrasonic Welding

Ultrasonic welding was discussed in detail in Chapter 6. There are specific guidelines for the design of energy directors or shear joints. Below are links to design guides on the subject from Branson Ultrasonics Corporation and Dukane:

http://www.emersonindustrial.com/en-US/documentcenter/BransonUltrasonics/Plastic%20Joining/Ultrasonics/Technical%20Info/PW-3_Part_Design_for_Ultrasonic_Welding_(single_pgs)_hr.pdf

https://www.dukane.com/us/Documents/DesignGuides/Guide%20To%20US%20Plastic%20Assembly.pdf

Round Energy Directors

More information from Dukane:

https://www.assemblymag.com/articles/93031-round-energy-directors-for-ultrasonic-welding

9.4 Tooling

9.4.1 Classes of Molds

The Society of Plastics Industry has created a document for the classification of molds, which provides guidelines for their construction based on the minimum number of cycles they will last for. This document can be downloaded from:

http://www.teampti.com/content/pdf/spi_guidelines.pdf

9.4.2 Tool Polish

The designer has to generally specify the finish on the mold to obtain the desired surface on the part–from mirror-like to matt (A1-D3). This standards document can be downloaded from the site below:

https://www.milacron.com/wp-content/uploads/2016/04/DME-Micro-Finish-Mold-Tool-Finish-Guide.pdf

9.4.3 Texture

A vast number of parts have a texture applied to the surfaces both for aesthetic appeal and wear. Although there are many suppliers across the world that perform textured surfaces on the molds, the most widely used ones are from Mold-Tech. Mold-Tech provides both the specifications and actual texture plaques for the designers to get a good idea of the appearance, texture depth, and draft requirements.

The specifications can be downloaded from:

http://www.mold-tech.com/downloads/Texture-Tech-Data.pdf

The "book" of textures can be ordered from:

http://www.mold-tech.com/purchase/

9.4.4 Total Tolerance

The total tolerance variation on a given part depends on the resin, design, tooling, and processing. Most of the guidelines provided by various resin suppliers and others are just that—guidelines. To have a tightly toleranced part, all of the above have to be very precisely controlled. In addition, "steel safe" conditions as described earlier have to be built into the tool to further fine tune the tolerances after an actual part has been molded using the optimized conditions—especially in the case of components such as precision snaps and gears.

The best tolerance values based on scientific reasoning that I have used are contained in DIN 160901 "Plastics Mouldings - Tolerances and Acceptance for Linear Dimensions." A link to this document is provided below:

https://www.beuth.de/en/standard/din-16742/192658755

9.4.5 Sprue, Gate, and Vent Design

While the mechanical details of molding tools need to be best left to the tool designers and manufacturers, proper cooling, sprue, gate, and vent design should be areas that the designer needs to be fully a part of. The cooling portion was discussed in detail in previous chapters. The design and mold guide from Bayer has excellent discussion on the other above-mentioned areas. Here is the link again:

https://www.plastics.covestro.com/Library/Technology-brochures.aspx

9.5 Processing

Scientific Molding

The chapter on processing (Chapter 5) discussed scientific molding. In my opinion, John Bozzelli and Suhas Kulkarni deserve a lot of credit in the promotion of this technique through their affordable and convenient training programs. They have a wealth of information at their sites with the links below:

John Bozzelli: http://www.scientificmolding.com/

Suhas Kulkarni: http://www.fimmtech.com/

Suhas also has an easy to understand book on the subject, *Robust Process Development and Scientific Molding*, published by Hanser, which I highly recommend:

http://www.hanser-elibrary.com/isbn/9783446422759

9.6 Simulation and Analysis

9.6.1 Simulation

Even if the designer does not have in-house simulation capabilities, it is highly recommended that simulation be performed on all but the very simple parts through the software suppliers, molders, or outside resources. This extra effort and cost will be paid back many fold in most cases, in terms of part robustness, cost, and time to market.

The three most used process simulation software are Moldflow, Moldex 3D, and Sigmasoft. Their links, respectively, are:

http://www.autodesk.com/products/moldflow/overview

http://www.moldex3d.com/en/

http://www.sigmasoft.de/

9.6.2 Design Checking and Design for Performance

Most organizations have traditionally used checklists, which in theory prevent errors in design. In reality, though, checklists are cumbersome to use because of the number of items that need to be checked. In most cases, they also point out errors "after the fact" and therefore when it may be too late to fix them.

An alternate is the software called DFMPro–a CAD integrated design assistant, which helps design engineers quickly verify the designs per the recommended design, with built-in industry best practices and recommended guidelines from various industry standard publications for improving product performance. Design engineers can quickly verify their plastic designs in real time for uniform wall thickness, draft angles, undercuts, thin steel conditions on mold, and various recommended design parameters for ribs, bosses, snap-fits, and many other common guidelines. See Figure 9.3 as an example of this analysis, which took just minutes. A trial version is available from the following link:

www.dfmpro.com

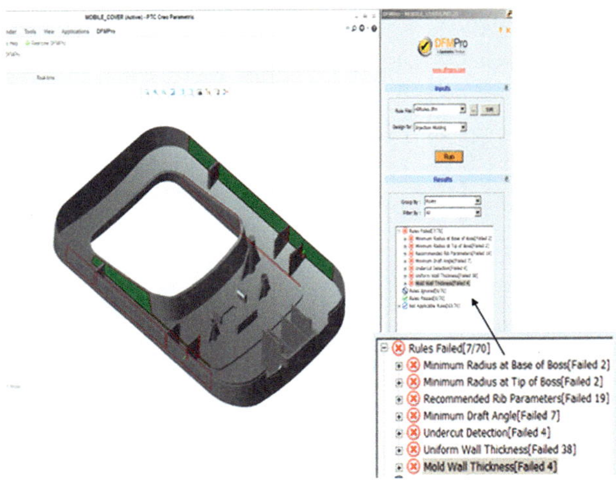

Figure 9.3 Typical DFMPro output highlighting design errors

■ 9.7 Failure Analysis Tools

Photoelastic Analysis

Polarized light passing through a stressed clear plastic part splits into two wave fronts traveling at different velocities, each parallel to a direction of principal stress but perpendicular to each other. This leads to the stressed material having two different indices of refraction, a property known as birefringence. (Adapted from Bayer literature per the link below.)

The result is different colors indicating different levels of stress as shown in Figure 9.4 (also in Chapter 3). This was obtained by placing the surface to be examined between two layers of polarizing sheets and passing light through them. Polarized sheets can be obtained from many suppliers. As an example, they can be obtained very inexpensively from:

http://www.spectrum-scientifics.com/

Where possible, it may be a good idea to mold a few parts in a clear resin such as polycarbonate or acrylic before the mold is textured, to proactively study the stress patterns.

Below is a very useful article, "Photoelastic Stress Analysis of Polycarbonate Medical Parts," by Mark Yeager at Covestro (formerly Bayer):

http://www.strainoptics.com/files/Photoelastic%20Stress%20Analysis%20of%20PolycarbonateMedical%20Parts.pdf

See also the following video:

https://www.youtube.com/watch?v=BdSaA9byKms

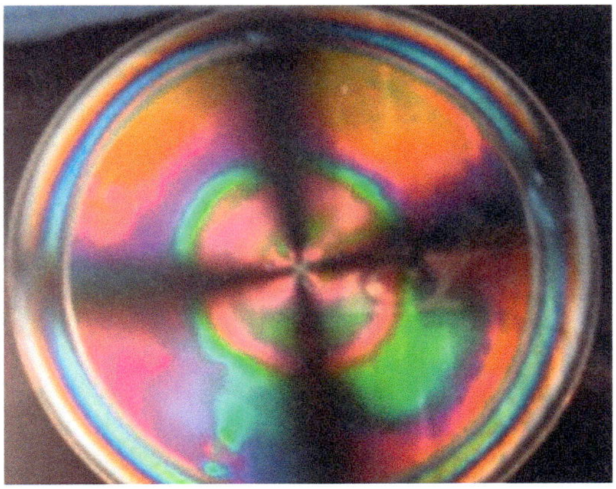

Figure 9.4 Birefringence pattern near the gate in a PS drinking cup

9.8 Six Sigma and Statistics

9.8.1 Basic Statistics

An understanding of basic statistics is of paramount importance to a practicing engineer. There are many sites and books available for this. One of the more simple books I have found for basic statistics and everyday reference is *Statistics for Six Sigma Made Easy* by Warren Brussee. See link below.

http://mhebooklibrary.com/doi/abs/10.1036/9780071502023

9.8.2 Design of Experiments (DOE)

As described earlier, the design of experiments is a systematic approach to finding the most critical factors affecting the quality of a process or product. A basic understanding of this is also of great advantage to practicing engineers.

There are many books to study the subject. Also, there are many software packages available to conduct the studies. Some of the ones that I have used with great success are:

- Minitab: It also has most of the leading statistical analysis tools in addition. A trial copy can be downloaded for free for 30 days.

  www.minitab.com
- Design Expert and Design Ease by Stat-Ease: Very useful and user-friendly software, mainly for design of experiments. A trial copy can also be downloaded free for 30 days.

  http://www.statease.com/
- An easy to use DOE module specializing in plastics is available from FIMMTECH. A free version is also available:

  http://www.fimmtech.com/

Statistics for Six Sigma Made Easy, mentioned above, also has an Excel-based approach for simple DOEs.

9.8.3 Process Capability

One of the ultimate aims for all design, tooling, and processing is to improve the process capability. The values related to process capability are expressed as Cp, Cpk, Pp, and Ppk. A good understanding of these terms and their calculations and significance is essential for any designer.

The following article, "Cp, Cpk, Pp and Ppk: Know How and When to Use Them," will provide the reader with a good grounding on the subject:

https://www.isixsigma.com/tools-templates/capability-indices-process-capability/cp-cpk-pp-and-ppk-know-how-and-when-use-them/

9.8.4 Six Sigma Tools

Statistics for Six Sigma Made Easy, above, is a very easy way to understand Six Sigma and the tools used within the methodology. Another very useful website is iSixSigma, with the following address:

https://www.isixsigma.com/

9.9 References

[1] This figure and associated discussion is adapted from UL IDES Prospector

[2] Adapted from *http://www.rohsguide.com/*

[3] *http://ec.europa.eu/environment/chemicals/reach/reach_en.htm*

[4] *http://ulstandards.ul.com/standard/?id=94_6*

[5] Covestro LLC

[6] *https://www.pinterest.com/eventcaddie/logo-printed-golf-balls/*

Index

E

F

G

Q

R

S

U

V

W

Y

Z